Chaotic
Dynamics of
Nonlinear Systems

Chaotic Dynamics of Nonlinear Systems

S. NEIL RASBAND

Department of Physics and Astronomy
Brigham Young University
Provo, Utah

WILEY

A Wiley-Interscience Publication

JOHN WILEY & SONS

New York · Chichester · Brisbane · Toronto · Singapore

Library of Congress Cataloging in Publication Data:
Rasband, S. Neil.
  Chaotic dynamics of nonlinear systems/S. Neil Rasband.
    p. cm.
  "A Wiley-Interscience publication."
  Bibliography: p.
  Includes index.
  ISBN 0-471-63418-2
  1. Chaotic behavior in systems. 2. Nonlinear theories.
I. Title.
  Q172.5.C45R37 1989
  003'.75—dc20   89-32903
                                           CIP

Printed in the United States of America

10 9 8 7 6 5 4 3 2

To Judi

PREFACE

In recent years the scientific community has witnessed the birth and initial development of a new paradigm for understanding complicated and seemingly unpredictable behavior. This new paradigm goes by the name of *Chaos*, referring to a scientific philosophy, an approach, and a set of methods to deal with manifestations of chaos in the physical sciences. To a large extent the enthusiasm that has developed for Chaos is the result of the breadth of its applications. These applications of Chaos for understanding complex and unpredictable behavior range across the spectrum of scientific disciplines. Indeed, Chaos has much the same flavor as classical thermodynamics in that the fundamental ideas and results seem applicable to a wide variety of different physical systems. There is probably no physical system exhibiting unpredictable behavior that is not presently being scrutinized through the lens of Chaos by someone.

Despite widespread interest and broad application, Chaos is a young science and as a consequence, the traditional examples are fewer and the standardization of methods is less well developed. With its roots in many areas of scientific inquiry, only in recent years have the examples and methods been welded into a new structure. In this textbook I have tried to introduce Chaos by presenting those topics and examples that seem to have risen to the top and become standard fare. However, for reasons of length choices must be made and the topics selected certainly reflect my own preferences. But I have tried to represent what most people in the community seem to feel are the important topics. Naturally, not all would agree on every point, and I would not expect anyone to agree with all of the selections or with the depth to which I have discussed them. Nevertheless, I believe that most of the "classical" topics in Chaos are represented.

To eliminate from the beginning any false expectations, I mention some topics that are not discussed: quantum chaos, noisy chaos, symbolic dynamics, and many, many examples of chaos occurring in specific physical systems. First and foremost, the book is intended to be useful as a textbook in a one-semester course taught in a physics department for seniors or first-year graduate students, and I have used this material for such a course. The audience, however, has not consisted only, or even primarily, of physics students. Certain topics and chapters require decidedly more background than others.

Chapter 8 on conservative dynamics expects the reader to be familiar with Hamiltonian dynamics. Sections 5.5 and 9.2 use some basic mathematical tools from differential geometry. However, the vast majority of the presentation depends only on some familiarity with differential equations and linear vector spaces. Even the reader with a limited knowledge of Hamiltonian dynamics or certain mathematical tools should be able to follow the presentation with only rarely a feeling of unfamiliarity.

The absolutely essential prerequiste the author expects the reader to bring to a study of this book is a willingness to do considerable numerical experimentation. The programming and numerical skills required for most of the examples are minimal, but a great deal of insight comes from personally performing numerical experiments on some of the classical problems in Chaos. Personal computers are adequate for doing everything in this book but, of course, may not suffice for tackling research problems in Chaos.

I wish to express my personal gratitude to colleagues who have fueled my interest by giving me copies of articles from diverse places and by generally encouraging me in this writing project. Particularly, I thank G. Mason, G. Hart, R. Shirts and E. Räuchle. I thank H. Stokes for frequent suggestions on TEX formatting and T. Knudsen for help in preparation of the manuscript. I especially thank my colleague and friend Ross Spencer for reading the manuscript and making literally hundreds of suggestions. The book is significantly better than it would otherwise be as a result of his help.

My deepest thanks go to my family since a considerable portion of the time necessary to complete this work has been taken from hours that rightfully belonged to them. The completion of this project would not have been possible without the love and support of my wife and children.

S. Neil Rasband

Provo, Utah
July 1989

CONTENTS

CHAPTER ONE

INTRODUCTION

*There are more things in heaven and earth,
Horatio, than are dreamt of in your philoso-
phy.*
(W. Shakespeare, Hamlet, Act I, Scene 5)

Arguably the most broad based revolution in the worldview of science in the twentieth century will be associated with chaotic dynamics. Yes, I know about Quantum Mechanics and Relativity, and for physicists and philosophers these theories must rank above Chaos for their impact on the way we view the world. My assertion, however, refers to science in general, not just to physics. Leaving improved diagnostic instrumentation aside, it is not clear that Quantum Mechanics or Relativity have had any appreciable effect whatever on medicine, biology, or geology. Yet chaotic dynamics is having an important impact in all of these fields, as well as many others, including chemistry and physics.

Surely part of the reason for this broad application is that chaotic dynamics is not something that is part of a specific physical model, limited in its application to one small area of science. But rather chaotic dynamics is a consequence of mathematics itself and hence appears in a broad range of physical systems. Thus, although the mathematical representations of these physical systems can be very different, they often share common properties. In this introductory chapter we outline in a qualitative way some of the common features of chaos and introduce the reader to some chaotic phenomena. We further introduce some of the methods employed in the study of chaotic dynamics. Precision is left to discussions in subsequent chapters.

1.1 Chaos and Nonlinearity

The very use of the word "chaos" implies some observation of a system, perhaps through some measurement, and that these observations or measurements vary unpredictably. We often say observations are chaotic when there is no discernable regularity or order. We may refer to spatial patterns as chaotic if they appear to have less symmetry than other, more ordered states. In more technical terms we would say that the correlation in observations separated by either space or time appears to be limited. However, from the

1

outset we must make clear that we are not speaking of the observation of random events, such as the flipping of a coin. Chaotic dynamics refers to *deterministic development* with chaotic outcome. Another way to say this is that from moment to moment the system is evolving in a deterministic way, i.e., the current state of a system depends on the previous state in a rigidly determined way. This is in contrast to a random system where the present observation has no causal connection to the previous one. The outcome of one coin toss does not depend in any way on the previous one. A system exhibiting chaotic dynamics evolves in a deterministic way, but measurements made on the system do not allow the prediction of the state of the system even moderately far into the future.

Whenever dynamical chaos is found, it is accompanied by *nonlinearity*. Nonlinearity in a system simply means that the measured values of the properties of a system in a later state depend in a complicated way on the measured values in an earlier state. By complicated we mean something other than just proportional to, differing by a constant, or some combination of these two. Although by these remarks, we do not mean to imply that somewhat complicated phenomena cannot be modeled by linear relations.

A simple, nonlinear, mathematical example would be for the observable x in the $(n+1)$th state to depend on the square of the observable x in the nth state, i.e., $x_{n+1} = x_n^2$. Such relations are termed *mappings*, and this is a simple example of a nonlinear map of the nth state to the $(n+1)$th state. A familiar physical example would be the temperature from one moment to the next as water is brought to a boil. At the end of this process the temperature in the $(n+1)$th state is just equal to the temperature in the nth state, but this is clearly not true as the water is being heated to its boiling temperature. Frequently the problem of modeling real-world systems with mathematical equations begins with a linear model. But when finer details or more accurate results are desired, additional nonlinear terms must be added.

Naturally, an uncountable variety of nonlinear relations is possible, depending perhaps on a multitude of parameters. These nonlinear relations are frequently encountered in the form of difference equations, mappings, differential equations, partial differential equations, integral equations, or even sometimes combinations of these. As we look deeper into specific causes of chaos, we shall see that chaos is not possible without nonlinearity. Nonlinear relations are not sufficient for chaos, but some form of nonlinearity is necessary for chaotic dynamics.

Having considered briefly nonlinear mappings, we now consider somewhat more closely systems modeled by differential equations. It is convenient when discussing the properties of differential equations to write them in a standard, first-order form:

$$\dot{\mathbf{x}} = \mathbf{f}(\mathbf{x}, t). \tag{1.1}$$

If the $\mathbf{f}$ in (1.1) is independent of t, then the equation is said to be *autonomous*; otherwise it is *nonautonomous*. For such a system to be chaotic it must have more than one degree of freedom, or be nonautonomous. We illustrate this with the familiar example of a simple pendulum. The differential equation for a simple pendulum is often written in the form

$$\ddot{x} + \omega_0^2 \sin x = 0, \tag{1.2}$$

where x represents the angular displacement of the pendulum from the vertical position, two overdots denote two derivatives with respect to time in the usual way, and ω_0 denotes the natural frequency of the pendulum for small angular displacements. Even though this system is highly nonlinear, it does not exhibit chaotic dynamics. There is only the single degree of freedom associated with x and the right-hand side is the constant zero. If, instead, we replaced the zero in (1.2) with some function $f(x,t)$, then the system becomes nonautonomous and may exhibit chaotic dynamics, depending of course, on the exact nature of the function $f(x,t)$. In effect the time t has become an additional degree of freedom.

To put the differential equation (1.2) in the standard form (1.1) and to make explicit the notion that time is a degree of freedom, we define a new independent variable θ, and a new dependent variable $y = dx/d\theta$. Then with the driving term $f(x,t)$ on the right, (1.2) becomes the system

$$\frac{dx}{d\theta} = y, \quad \frac{dy}{d\theta} = -\omega_0^2 \sin x + f(x,t), \quad \frac{dt}{d\theta} = 1.$$

In this form the system consists of three, first-order differential equations and is nonautonomous. Frequently, such a system is said to have $1\frac{1}{2}$ degrees of freedom, since very often dynamical systems, particularly those resulting from Hamiltonian mechanics, have a pair of equations for every degree of freedom.

Although simple quadratic maps and forced, nonlinear oscillators like the preceeding examples may not appear to offer much promise for displaying a rich diversity of chaos, the opposite is true. We will see that indeed within these very simple nonlinearities lurk the seed of nearly all chaotic phenomena, and the bulk of this work is devoted to the study of such simple systems.

One of our major objectives is to classify and characterize deterministic systems exhibiting chaotic dynamics. Thus our characterization of nonlinearity as an essential ingredient for chaotic dynamics marks the beginning of this classification effort. We have further pointed out that for a system with one degree of freedom the differential equation must be nonautonomous. We now illustrate these points and the development of chaos with the familiar example of a simple harmonic oscillator.

1.2 The Kicked Harmonic Oscillator

To introduce many of the concepts and ideas that will be studied in subsequent chapters, we study the motion of a simple harmonic oscillator subject to

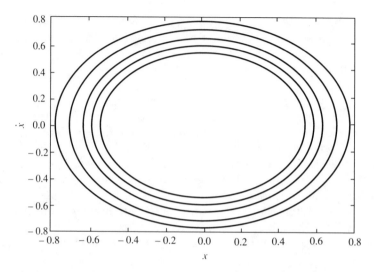

FIGURE 1.1   Sample phase-plane trajectories for the simple harmonic oscillator without kicks.

a periodic impulse. We refer to this system as the *kicked harmonic oscillator*. The equation of this system is given by

$$\ddot{x} + \omega_0^2 x = A f(x) \sum_{n=1}^{\infty} \delta(t - nT), \qquad (1.3)$$

where ω_0 is the natural frequency of the oscillator, A is the amplitude of the kicks, and $f(x)$ is an arbitrary function of x, but not of t. Figure 1.1 shows the familiar phase-plane trajectories for the case where $A = 0$, i.e., the harmonic oscillator without kicks. Each ellipse corresponds to a fixed value of the energy of the oscillator. With $A \neq 0$, the right-hand side of (1.3) depends on time t; this differential equation is therefore nonautonomous.

In an interval between kicks the right-hand side of (1.3) is zero, and the solution is familiar:

$$x(t) = A_k \cos \omega_0 t + B_k \sin \omega_0 t, \qquad (k-1)T < t < kT, \qquad (1.4)$$

and

$$\dot{x}(t) = -\omega_0 A_k \sin \omega_0 t + \omega_0 B_k \cos \omega_0 t, \qquad (1.5)$$

where $k = 1, 2, \ldots$. For each k, at $t = kT$ we demand that the position of the one-dimensional oscillator be continuous but that the velocity (momentum) change discontinuously. This discontinuous change in the velocity is computed by integrating (1.3) from $(kT - \epsilon)$ to $(kT + \epsilon)$ and then taking the limit as $\epsilon \to 0$. We find easily the following relationship between the coefficients in the k interval and those in the $(k + 1)$ interval.

$$A_{k+1} = A_k - \frac{A}{\omega_0} f(x_k) \sin \omega_0 kT, \qquad (1.6)$$

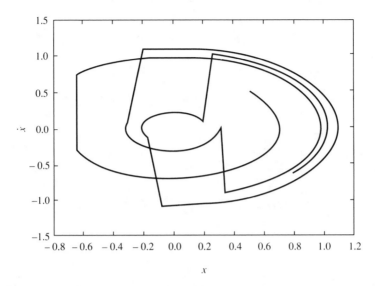

FIGURE 1.2 A section of a phase-space trajectory for a linear kicked oscillator. The discontinuous jumps in $\dot{x}$ are a result of the kicks.

$$B_{k+1} = B_k + \frac{A}{\omega_0} f(x_k) \cos \omega_0 kT, \tag{1.7}$$

where

$$x_k = A_k \cos \omega_0 kT + B_k \sin \omega_0 kT,$$
$$\dot{x}_k = -\omega_0 A_k \sin \omega_0 kT + \omega_0 B_k \cos \omega_0 kT. \tag{1.8}$$

The subscript k on x and $\dot{x}$ refer to a time infinitesimally prior to the kick at kT. Using (1.8) with (1.6) and (1.7), plus a little algebra, yields the relation

$$\begin{pmatrix} x_{k+1} \\ \dot{x}_{k+1} \end{pmatrix} = \begin{pmatrix} \cos \omega_0 T & \omega_0^{-1} \sin \omega_0 T \\ -\omega_0 \sin \omega_0 T & \cos \omega_0 T \end{pmatrix} \begin{pmatrix} x_k \\ \dot{x}_k + A f(x_k) \end{pmatrix}, \tag{1.9}$$

which gives the position and velocity just before the time $(k + 1)T$ in terms of the position and velocity just before the time kT.

The relationship between the coefficients in the k interval to those in the $(k + 1)$ interval is an example of a two-dimensional mapping. Choosing the driving term in (1.3) to be a sum of delta functions is the feature that allows us to obtain the solution to the differential equation for the kicked harmonic oscillator in terms of the mapping represented by (1.9). The nonlinearities are introduced by the choice made for $f(x)$. With A and $f(x)$ not equal to zero, the system is nonautonomous and thus equivalent to more than one degree of freedom.

For $f(x) = 1$ or x, the mapping (1.9) is linear and invertible. In light of our previous remarks, no chaotic dynamics is to be expected. Such a case is, however, still nonautonomous — just not nonlinear. A plot of a segment of a phase-space trajectory for $f(x) = 1$ is given in Fig. 1.2. The trajectory crossings are a consequence of the time dependent driving term but can be eliminated by plotting the trajectory in extended phase space as in Fig. 1.3.

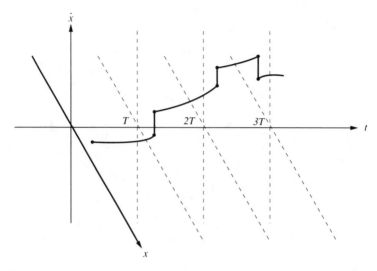

FIGURE 1.3   A sketch of a possible trajectory in extended phase space for the kicked harmonic oscillator. The kicks, and consequently discontinuous jumps in the velocity, occur at $t = T, 2T, \ldots$.

From (1.6) and (1.7) with $f = 1$ we obtain immediately

$$
A_k = A_1 - \frac{A}{\omega_0} \sum_{n=1}^{k-1} \sin(2\pi n \omega_0 / \Omega),
$$

$$
B_k = B_1 + \frac{A}{\omega_0} \sum_{n=1}^{k-1} \cos(2\pi n \omega_0 / \Omega),
$$

$$(1.10)$$

with $k = 2, 3, \ldots$ and $\Omega = 2\pi/T$. If $\omega_0 = \Omega$, i.e., if the kicks come at a frequency equal to the natural frequency of the oscillator, the coefficient $B_k \to \infty$ with k. The velocity and hence the energy of the oscillator become unbounded. This situation is called *resonance*. Resonance is a phenomenon occurring in a great many nonlinear systems leading to the destruction of the integrable behavior. The issue of resonance will reappear often in subsequent sections as we consider dynamics of nonlinear systems.

For $\omega_0 \neq \Omega$ the series in (1.10) can be summed to give

$$
A_k = A_1 + \frac{A}{\omega_0} \sin \pi k \left(\frac{\omega_0}{\Omega}\right) \left[\cos \pi k \left(\frac{\omega_0}{\Omega}\right) - \cot \pi \left(\frac{\omega_0}{\Omega}\right) \sin \pi k \left(\frac{\omega_0}{\Omega}\right)\right]
$$

$$
B_k = B_1 + \frac{A}{\omega_0} \left[\sin \pi k \left(\frac{\omega_0}{\Omega}\right) \left[\sin \pi k \left(\frac{\omega_0}{\Omega}\right) + \cot \pi \left(\frac{\omega_0}{\Omega}\right) \cos \pi k \left(\frac{\omega_0}{\Omega}\right)\right] - 1\right]
$$

$$(1.11)$$

If the ratio (ω_0/Ω) is a rational number, then there will always exist some k for which A_k and B_k return to their inital values, and the system is periodic.

As an alternative to a trajectory plot in extended phase space, which becomes impractical after a few periods, it is convenient to study the time

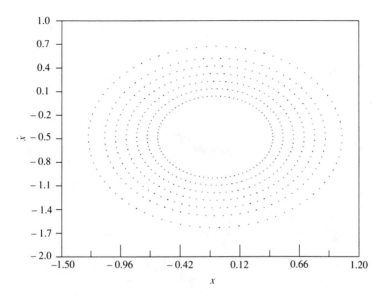

FIGURE 1.4   Poincaré Section plot for a kicked harmonic oscillator but with the driving term independent of x.

evolution of this system by making a point in the $(x, \dot{x})$ phase plane at the values of $t = T, 2T, \ldots$, i.e., at values of t corresponding to multiples of the period of the driving function. Such a plot for a dynamic system is called a *Poincaré section*. Figure 1.4 is a Poincaré section plot for the system represented by (1.10), (1.11) and we see that the phase points always lie on ellipses, just as for the oscillator without kicks.

Comparing an orbit in Fig. 1.1 with the orbit in Fig. 1.2 dramatically demonstrates that a linear, time-dependent driving term alters the orbits in phase space. But this change in the nature of the phase-space orbits still does not go so far as to produce any chaotic dynamics. The relation between the (A_k, B_k) and (A_{k+1}, B_{k+1}) is still linear in (1.10) and (1.11). Nonlinearity is still absent in the system producing Fig. 1.4. Exercise 1.3 considers the same issues with $f(x) = x$.

We now change from $f(x) = 1$ to $f(x) = x^4$ and examine the Poincaré section plots for orbits with initial conditions similar to those producing the plots of Fig. 1.4. The Poincaré sections now produce Fig. 1.5, which is quite different from Fig. 1.4. We see two highly distorted elliptical orbits, an inner and an outer one, enclosing a seven-period island chain. Around the outer edge of this island chain there is a small, but finite, layer of chaotic orbits. The centers of the islands are called O points and the points between, joining the individual "islands," are called hyperbolic or X points. The insert in the center of Fig. 1.5 shows a magnified view of the intersection points of a *single* orbit in the neighborhood of the indicated X point. The reader should bear in mind that the insert only shows one of the seven X points, all of which are

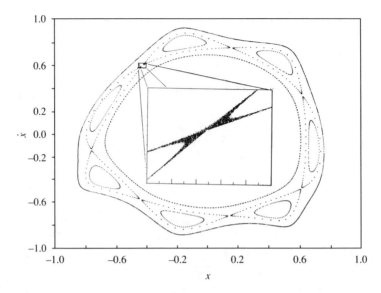

FIGURE 1.5   Poincaré section plot for a kicked harmonic oscillator with a dependence
of the form x^4 in the drive. The intersection points of a single trajectory produce the points
forming the island chain. The same is true for the outer closed curve, the inner closed curve,
and the chaotic region magnified in the insert.

connected by a thin chaotic layer around the island structure. The chaotic
region occupies a small but finite region in the phase plane. One of the most
characteristic features of chaotic dynamics can be seen by considering two
trajectories in the chaotic region that have nearly identical initial conditions.
After a finite number of iterations of the map, the intersection point for one
trajectory is completely unrelated to the intersection point for the second
trajectory. This is our first example of chaotic behavior from deterministic
dynamics. This feature is commonly referred to as *sensitive dependence on
initial conditions.* Despite sensitive dependence on initial conditions and numerical roundoff, Hammel et al. (1988) have shown that the computation of
chaotic orbits for a large number of periods as in Fig. 1.5 is still meaningful.

These few examples, and the kicked harmonic oscillator in particular,
have illustrated the necessity for nonlinearity in producing chaotic dynamics.
We further illustrated how Poincaré sections can be a useful tool in displaying
chaotic consequences. For the kicked harmonic oscillator it was possible to
obtain a mapping to advance the system in time, and it should be clear that
this is much easier than the numerical integration of a system of differential
equations. Partly because maps are easier to advance, and partly because of
the importance of Poincaré section maps, considerable attention is devoted
to mappings in subsequent chapters. This begins in the next chapter with a
study of one-dimensional maps where we also develop additional methods to
supplement Poincaré plots for studying and recognizing chaotic behavior.

1.3 Examples

The following is a selected list of some situtations where chaotic dynamics is manifest or appears to play a role.

1. Turbulence is believed to be the classic example of a system evolving deterministically, yet exhibiting chaotic behavior. Transistions to turbulence in Couette flow have been studied by Swinney and Gollub (1978), Swinney (1983, 1985), and Brandstater and Swinney (1987).

2. Thermal convection in fluids, called Rayleigh-Bernard convection, provides another example of transistion to turbulence. This has been one of the most fruitful applications experimentally and theoretically. It was in this system that chaotic dynamics was first appreciated theoretically with the work of Lorenz (1963). The Lorenz model is of such importance historically, and there has been so much work done on it, that the Lorenz equations have become one of the important examples for chaotic dynamics. Experimentally, it was careful measurements on liquid helium confined in a cell heated from below that led to stunning experimental confirmation of some of the predictions of chaotic dynamics by Libchaber and Maurer (1978). The Lorenz equations are considered in Chapter 6.

3. Supersonic panel flutter, important for supersonic aircraft and rockets, was studied by Kobayashi (1962).

4. Some chemical reactions, and in particular the Belousof-Zhabotinsky reaction, exhibit chaotic dynamics as discussed by Roux (1983) and Epstein (1983).

5. Optically bistable laser cavities have been studied by Ikeda et al. (1980) and Gibbs et al. (1981). Atmanspacher and Scheingraber (1986) have investigated several measures of chaos in a continuous-wave dye laser.

6. Cardiac dysrhythmias, or abnormal cardiac rhythms, have been discussed by Glass et al. (1983). In addition to the dynamics of the heart, its very structure has several manifestations of self-similar geometrical structures called fractals. Fractal structures are commonly the result of nonlinear dynamics, and, although the dynamics governing growth and development of the heart are unknown, fractal structures are detailed in the vascular network for the heart. Furthermore the cardiac impulse itself is transmitted to the ventricles via an irregular fractal network. Many such fractal structures in physiology are reviewed by West and Goldberger (1987).

7. There are many examples of nonlinear electrical circuits that exhibit chaotic dynamics. One famous example that has for many decades provided a model for nonlinear vibrations is the oscillator described by Van der Pol and Van der Mark (1927). Nonlinear circuits have provided analog devices for modeling many of the nonlinear equations discovered in one context or another.

8. Ecology and biological population dynamics provide a simple and instructive example of a dynamical system exhibiting chaotic dynamics. This example comes to us under the name of the "logistic equation." This equation may describe the variations in nonoverlapping biological populations from one year to the next. This equation and its importance were pointed out in an early review by May (1976). We study this classic example in the next two chapters.

9. Vibrations of buckled elastic systems have provided experimental examples of double-well potential systems. These systems have been studied experimentally and theoretically by Moon and Holmes (1979, 1980) and Holmes and Whitley (1983) as realizations of Duffing's equation, which is also one of the classical systems studied in nonlinear oscillations.

10. Chaotic dynamo models have been proposed for representing the geomagnetic field reversals and have been studied by Cook and Roberts (1970). A review has been given by Bullard (1978). We study this example in detail in Chapter 6.

11. Several types of standard chaotic behavior have been observed in simple plasma systems and reported in Cheung and Wong (1987) and Cheung et al. (1988).

12. A number of simple experiments suitable for classroom demonstration of chaos have been described by Briggs (1987).

13. Several researchers have claimed that EEG data suggests that chaotic neural activity plays a role in the processing of information by the brain. See Harth (1983), Nicolis (1984), and Skarda and Freeman (1987).

14. By constructing a special computer for the single purpose of studying the stability of planetary orbits over long time scales Sussman and Wisdom (reported by Lewin, 1988) have found the orbit of Pluto to be chaotic on a time scale of about 20 million years.

This selected list of examples illustrates the broad range of scientific investigation that has been affected by studies in chaotic dynamics. I offer to the reader the personal challenge to find some previously unmentioned system in nature exhibiting chaotic behavior. Chaos can make life interesting in many ways.

Exercises

1.1 Consider a kicked rotor with its dynamics modeled by the equation

$$\ddot{\phi} + \gamma\dot{\phi} = Af(\phi) \sum_{n=1}^{\infty} \delta(t - nT),$$

where ϕ is the angle of the rotor, measured from some fiducial point, and γ is the damping constant (Schuster, 1984). If $\phi_n(t)$ is the solution for

$(n-1)T < t < nT$, show that

$$\phi_{n+1} = \phi_n + \frac{1 - e^{-\gamma T}}{\gamma}\left(\dot{\phi}_n + Af(\phi_n)\right),$$

$$\dot{\phi}_{n+1} = e^{-\gamma T}\left(\dot{\phi}_n + Af(\phi_n)\right),$$

where $\phi_n = \phi_n(nT)$ and $\dot{\phi}_n = \dot{\phi}_n(nT)$. For some choice of the parameters A and γ, and for a choice of $f(x)$, give an example of a map exhibiting chaos.

1.2 Obtain the so-called "standard map" (cf. Section 8.5) in the form

$$\phi_{n+1} = \phi_n + y_{n+1}$$
$$y_{n+1} = y_n + k\sin\phi_n,$$

by considering the nondissipative limit $\gamma \to 0$ in the previous exercise. Define a scaled velocity variable and make an appropriate choice for the function $f(\phi_n)$.

1.3 Consider a kicked harmonic oscillator as given in (1.3). Choose $f(x) = x$ and for several choices of A and the ratio (ω_0/Ω) show that one either gets unbounded motion (resonance) or regular ellipses in a Poincaré section.

1.4 Show explicitly that Eqs. (1.9) with $\omega_0 = 1$ and $f(x) = 1$ correspond to the simple rotation of a vector in phase space.

CHAPTER TWO

ONE-DIMENSIONAL MAPS

*Simple dynamical systems do not necessarily
lead to simple dynamical behavior.*
(R. M. May, 1976)

In the previous chapter some basic concepts in chaotic dynamics were introduced by considering a kicked harmonic oscillator. As a consequence of the
delta-function nature of the kicks, the dynamics could be formulated in the
form of a two-dimensional map in phase space. However, as we proceed from
an introduction to a systematic study, we are well advised to retreat to a
consideration of one-dimensional maps before tackling the complexity of two
dimensions.

We write such one-dimensional maps in one of the two general forms:

$$x_{n+1} = f(x_n) \quad \text{or} \quad x' = f(x), \tag{2.1}$$

where x_{n+1} or x' denotes the value of the variable x after application of the
map. In this chapter two examples are considered in detail. The first example
considered is the *tent map*:

$$\Delta_\mu(x) = \mu\left(1 - 2\left|x - \tfrac{1}{2}\right|\right) = 2\mu \begin{cases} x, & \text{if } 0 \le x \le \tfrac{1}{2}; \\ 1 - x, & \text{if } \tfrac{1}{2} \le x \le 1. \end{cases} \tag{2.2}$$

The second example is the logistic map:

$$F_\mu(x) = \mu x(1 - x), \tag{2.3}$$

or often written

$$x_{k+1} = \mu x_k(1 - x_k). \tag{2.4}$$

In both cases μ is a parameter that dramatically affects the behavior of the
map, as we shall see shortly. Figure 2.1 is a plot of these two maps for specific
choices of the map parameter μ.

The tent map serves as our first example for studying chaos and the
logistic map as our most important example.

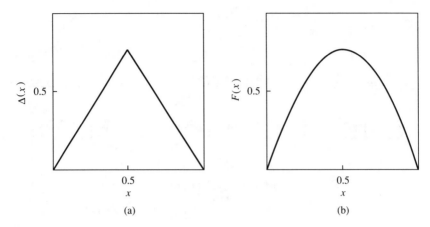

FIGURE 2.1   Plot of the tent map (a) and the logistic map (b) with the map parameter $\mu = 3/4$ and $\mu = 3$, respectively.

2.1 The Tent Map

The tent map $\Delta_\mu(x)$ defined in (2.2) is sketched in Fig. 2.2 for two choices of the map parameter μ. The dynamics of this map, as with the other one-dimensional maps we consider, is made transparent by using graphical analysis. This is particularly so when repeated application of the map is desired. Recall that each application of the map corresponds to an advance in time when such maps model dynamics. Repeated iterations of the map are easily done graphically by the following procedure: Given an x_0 value on the horizontal axis, follow a vertical line (constant x) until it intersects the curve $\Delta_\mu(x)$ at $\Delta_\mu(x_0)$. From this intersection point on the curve $\Delta_\mu(x)$ follow a horizontal line (constant y) until it intersects the line $y = x$. From this intersection point on the line $y = x$ follow a vertical line until it intersects the curve $\Delta_\mu(x)$ again. This then gives the value of two iterations of the map. Figure 2.2 sketches a few iterations of the map $\Delta_\mu(x)$.

One thing apparent from the graph is that if μ is large enough there is a value of $x \neq 0$, call it x_*, where the curves $y = x$ and $y = \Delta_\mu(x)$ are equal. In general a point satisfying

$$x_f = f(x_f) \qquad (2.5)$$

is called a *fixed point* and is denoted as x_f. The point $x_0 = 0$ is a fixed point for $\Delta_\mu(x)$ and satisfies (2.5). The nonzero fixed point satisfying (2.5) for the tent map is easily found to be

$$x_* = \frac{2\mu}{1 + 2\mu}, \qquad \mu > \frac{1}{2}. \qquad (2.6)$$

Not all fixed points have identical properties and one important characterization of fixed points is their stability. Let us consider the stability of the

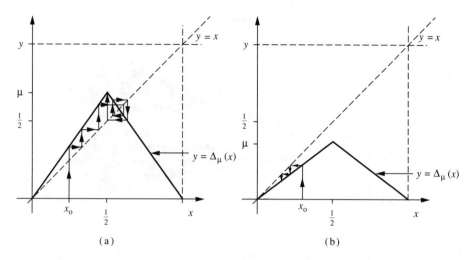

FIGURE 2.2   Sketches of the tent map $\Delta_\mu(x)$ for two choices of μ with several iterations of the map according to graphical analysis shown.

fixed point x_* by considering the distance of successive iterates from x_*. If $\mu > \frac{1}{2}$, the map $\Delta_\mu(x)$ is clearly analytic in some small neighborhood U of x_*. Let the point $x_n \in U$ and let $\delta_n = x_n - x_*$. Then

$$|\delta_{n+1}| = |x_{n+1} - x_*| = |\Delta_\mu(x_n) - x_*| = |\Delta_\mu(x_* + \delta_n) - x_*|$$
$$= \left|\Delta_\mu(x_*) + \delta_n \frac{d\Delta_\mu(x)}{dx}\right|_{x_*} - x_*\right| = \left|\frac{d\Delta_\mu(x)}{dx}\right|_{x_*}\left||\delta_n|. \qquad (2.7)$$

Clearly $|\delta_{n+1}| < |\delta_n|$ if and only if $|d\Delta_\mu(x_*)/dx| < 1$. In arriving at this conclusion from (2.7), only the analyticity of the tent map near x_* was used. In general for a map $f(x)$, analytic in the neighborhood of a fixed point x_f, the map $f(x)$ is stable at x_f if and only if

$$\left|\frac{df}{dx}\right|_{x_f}\right| < 1. \qquad (2.8)$$

If $|df/dx|_{x_f} \neq 1$, then x_f is referred to as a hyperbolic fixed point.

Applied to the tent map at $x = 0$, we find $d\Delta_\mu(x)/dx = 2\mu$ and so the fixed point $x = 0$ is stable as long as $\mu < \frac{1}{2}$. For $\mu > \frac{1}{2}$, $x = 0$ is an unstable fixed point, and for $\mu = \frac{1}{2}$ we say it is marginally stable. The case $\mu = \frac{1}{2}$ is clearly a degenerate case since all points in the interval $[0, \frac{1}{2}]$ are then fixed points.

For $\mu > \frac{1}{2}$ there are two fixed points 0 and x_* with $\frac{1}{2} < x_* < 1$. For $0 \leq x < \frac{1}{2}$, $d\Delta_\mu/dx = 2\mu$; for $\frac{1}{2} < x \leq 1$, $d\Delta_\mu/dx = -2\mu$. Since $\mu > \frac{1}{2}$ for x_* to exist, it is clear that this fixed point is always unstable. For $\mu < \frac{1}{2}$, $x = 0$ is an attracting fixed point, as is easily seen from graphical analysis (cf. Fig. 2.2b). It is readily verified (cf. Fig. 2.2a), that for $\mu > \frac{1}{2}$ neither fixed point is attracting. Figure 2.2a shows that points near 0 move away until they get near x_*, then they move away from x_*, then away from 0, etc.

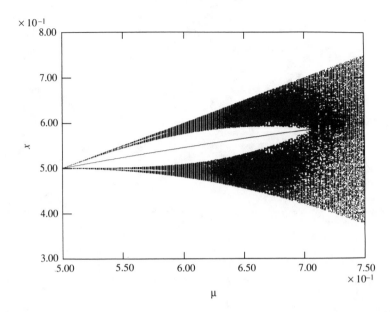

FIGURE 2.3   Iterates of the tent map $\Delta_\mu(x)$ as a function of μ with a single initial point x_0, which is chosen arbitrarily. For each value of μ an initial transient of 200 points is discarded. The solid curve gives the fixed-point x_*.

As a further look at this issue for a range of μ values consider Fig. 2.3. An arbitrary initial value x_0 can be chosen, but the iterates all end up on an attracting set. For values of μ near 0.5, the attracting set looks like two points in the resolution of the figure. For larger values of μ it is clear that the attractor consists of sets of points. It is further clear from Fig. 2.3 that the fixed point $x_f = x_*$ is not in the attractor. Certainly for values of μ approaching $\mu = 0.75$ the orbit of a single initial input point appears distributed all over a point set that at least appears dense in some finite interval. As $\mu \to 1$ this interval goes to (0,1). These maps $\Delta_\mu(x)$, at least for $\mu \simeq 0.75$, have every appearance of being chaotic.

We examine in greater detail the specific map for $\mu = 1$. For brevity we denote this map simply as $\Delta(x)$ and note that it is representative for $\mu \simeq \frac{3}{4}$. As is customary, we denote the iterates of a map with superscripts, e.g., $\Delta^2(x) = \Delta(\Delta(x)) = \Delta \circ \Delta(x)$, where the latter notation is usual for functional composition. The notation $\Delta^n(x)$ does not mean $\Delta(x)$ raised to the nth power, but rather

$$\Delta^n(x) \equiv \Delta \circ \Delta \circ \cdots \circ \Delta(x), \qquad (2.9)$$

with n occurrences of Δ. Consider Fig. 2.4a showing Δ^2 and Fig. 2.4b showing Δ^n.

Examination of Fig. 2.4b shows that if two nearby points are separated by a distance $\Delta x = \epsilon$, then after n applications of Δ we find $\Delta x = \epsilon 2^n$, which we obtain immediately since the slope of the linear pieces of Δ^n is $\pm 2^n$.

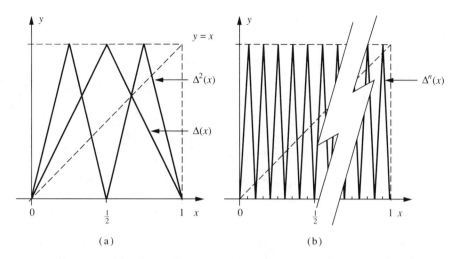

FIGURE 2.4   Sketch of the tent map Δ and its iterates Δ^2 and Δ^n.

When $\Delta x = 1$ the separation is as large as the interval itself, and this occurs after n iterations, where $n \simeq \ln(1/\epsilon)/\ln 2$. We find that even if ϵ is as small as 10^{-12}, after about 40 iterations there is no correlation between the image points. This is clearly a sensitive dependence on initial conditions—one of the most important characteristics of chaotic dynamics. In the following section we return to this exponential separation of nearby points and introduce the Lyapunov characteristic exponent as a measure of the separation.

One of the best ways to understand the appearance of chaos in noninvertible maps like the tent map is to think of them as consisting of two parts: (i) a stretching and (ii) a folding back. Stretching is produced by having the slope of the map be greater than one in magnitude. The folding back is produced by the decreasing part of the map on the interval $(\frac{1}{2},1]$. The stretching feature of the map produces sensitive dependence on initial conditions by making the iterated images of neighboring points diverge from each other. The folding back feature, which is required to keep the points confined to a bounded interval, mixes up these images making chaos possible. Such maps are not one-to-one and therefore not invertible. Linear maps are always invertible, and so we see once again the need for nonlinearity. Note that although the tent map is piecewise linear, it is still not invertible.

The stretching and folding properties of the tent map $\Delta_\mu(x)$ are illustrated in Fig. 2.5 for a μ in the range $\frac{1}{2} < \mu < 1$. We see the stretching that results because the slope is greater than 1. Consequently, points near 0 all get stretched out toward μ, but no points get mapped back into the interval $[0, 2\mu(1 - \mu)]$. Thus all points on the unit interval eventually get trapped in the interval $[2\mu(1 - \mu), \mu]$. The interval $[\mu,1]$ is mapped back into $[0,\mu]$ as depicted in Fig. 2.5a, but no points get mapped back into $[\mu,1]$. The behavior of the tent map on these intervals is clearly evident in Fig. 2.3.

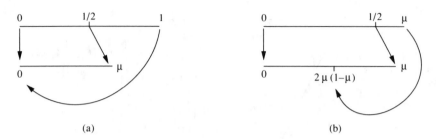

FIGURE 2.5   The tent map $\Delta_\mu(x)$ applied to (a) the interval [0,1] and (b) [0,μ].

Let us close this section by summarizing those things we have learned about chaos from the tent map.

1. Whether one has chaotic dynamics or not can depend on parameters characterizing the map, such as the parameter μ in the tent map $\Delta_\mu(x)$.
2. The fixed points are an important feature of maps and can be either repelling or attracting depending on the derivative of the function at the fixed point.
3. The sensitive dependence on initial conditions is a result of a stretching and folding operation. Such maps are noninvertible.

2.2 The Lyapunov Exponent in One Dimension

The exponential stretching that we are able to exhibit in detail for the tent map $\Delta_\mu(x)$ is conveniently expressed by the Lyapunov exponent $\lambda(x_0)$. In general the value of the Lyapunov exponent depends on the initial point x_0. In a later chapter we study Lyapunov exponents for multidimensional dynamical systems.

Consider two points x_0 and $x_0 + \epsilon$ mapped by the function $f: I \to I$, where $I \subset I\!R$ is some bounded interval on the real line $I\!R$. For n iterations of this map the Lyapunov exponent $\lambda(x_0)$ approximately satisfies the equation

$$\epsilon e^{n\lambda(x_0)} = f^n(x_0 + \epsilon) - f^n(x_0).$$

Dividing by ϵ and taking the limit as $\epsilon \to 0$ gives

$$e^{n\lambda(x_0)} = \frac{df^n(x)}{dx}\bigg|_{x_0}.$$

If we now take the limit as $n \to \infty$, then we have the definition for the Lyapunov exponent:

$$\lambda(x_0) \equiv \lim_{n \to \infty} \frac{1}{n} \ln\left|\frac{df^n(x)}{dx}\bigg|_{x_0}\right|. \qquad (2.10)$$

We see from these formulas that $\lambda(x_0)$ represents the average exponential stretching of initially nearby points. If we denote $x_i \equiv f^i(x_0)$ and recall (2.9)

$$f^n(x_0) = f(f(\cdots(f(x_0))\cdots)), \qquad (2.11)$$

then

$$\frac{df^n}{dx}\bigg|_{x_0} = \frac{df}{dx}\bigg|_{x_{n-1}} \frac{df}{dx}\bigg|_{x_{n-2}} \cdots \frac{df}{dx}\bigg|_{x_0} \equiv f'(x_{n-1})f'(x_{n-2})\cdots f'(x_0). \quad (2.12)$$

Consequently, Eq. (2.10) can be written in the form

$$\lambda(x_0) = \lim_{n\to\infty} \frac{1}{n} \sum_{i=0}^{n-1} \ln\left|f'(x_i)\right|. \qquad (2.13)$$

As an example consider the tent map $\Delta_\mu(x)$. For $\mu = 1$ we have $|\Delta'(x)| = 2$ for all $x \in [0,1]$, and thus $\lambda(x) = \ln 2$. This is consistent with the result of the previous section, where we found

$$f^n(x_0 + \epsilon) - f^n(x_0) = \Delta x = \epsilon 2^n = \epsilon e^{n\lambda}.$$

Canceling the common factor in the last equality and taking the natural logrithm of both sides gives $\lambda = \ln 2$. For arbitrary μ we find $\lambda = \ln 2\mu$, which is consistent with the result of Exercise 2.1 and shows exponential divergence of orbits for the tent map as long as $\mu > \frac{1}{2}$.

2.3 The Logistic Map

Let us now turn our attention to the second, and most important, example of this chapter: the logistic map. There is no better place to begin our investigation than by looking for those features summarized at the end of Section 2.1 for the tent map. We denote the logistic map in the form

$$F_\mu(x) = \mu x(1 - x) \qquad (2.14)$$

and in Fig. 2.6 we see some examples with various choices of the parameter μ. For values of $\mu > 4$ the map in (2.14) does not map the interval [0,1] into itself.

In Fig. 2.6 we also see several iterations of the maps for an initial point taken in the interval [0.0,0.25], using graphical analysis. The first two maps suggest that points converge to the fixed point under iteration. Graphical analysis in Fig. 2.6c, however, suggests that the iterates do not converge to the fixed point but map back and forth between two points, one on each side of the fixed point. The analysis that follows helps us understand the behavior of this map for various values of μ more clearly.

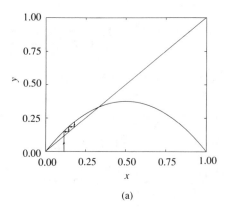

(a)

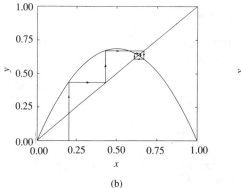

(b)

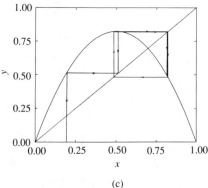

(c)

FIGURE 2.6   Some examples of the logistic map with various choices of μ. A few orbits are also followed by graphical analysis suggesting stability or instability for fixed points.

Solving the fixed-point equation

$$x_f = F_\mu(x_f), \qquad (2.15)$$

we find the fixed points 0 and x_*, where $x_* = (\mu - 1)/\mu$. It is clear that for x_* to be in the interval [0,1], $\mu \geq 1$. We find also that $F'_\mu(0) = \mu$ and $F'_\mu(x_*) = 2 - \mu$. From these simple results we learn that $x = 0$ is a fixed point that becomes unstable just as a second fixed point x_* is born. The fixed point x_* is stable for $1 < \mu < 3$. What happens when both fixed points become unstable for $\mu > 3$? We might expect something like what happened for the tent map $\Delta_\mu(x)$ with $\mu > \frac{1}{2}$. To explore this issue let us again numerically compute the iterates of the map as a function of the parameter μ.

With $0 < \mu < 1$ the point $x = 0$ is an attracting fixed point, and all points map into 0 under repeated application of F_μ. For $1 \leq \mu \leq 3$ all points go to x_*. Consequently, we begin plotting the attracting set of $F_\mu(x)$ as a function of μ with a value of $\mu = 2.8$ where the fixed point x_* is still an attracting fixed point. The result of these numerical computations is displayed in Fig. 2.7 and Plate I.

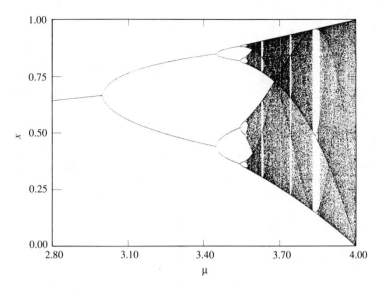

FIGURE 2.7   Iterates of the logistic map as a function of μ for $2.8 < \mu < 4.0$. An initial transient of 200 points has been discarded in each case.

Although there are similarities to Fig. 2.3, there are also striking differences. Perhaps the most notable feature is pitchfork-like bifurcations in the attractor. Initially the attracting set consists of a single point that bifurcates into two points at $\mu = 3.0$. Subsequently, these points bifurcate again into four points, which bifurcate into eight, and so on. The interval in μ between bifurcations decreases until eventually what looks like a chaotic set appears. The chaotic region appears interspersed with bands. In the bands only a small number of points appear to form the attractor.

In Section 2.2 we learned that the Lyapunov exponent λ was a good indicator of chaos, measuring the exponential separation for points initially nearby as the map is iterated. Figure 2.8 shows the Lyapunov characteristic exponent for points on the unit interval mapped under the logistic map as a function of μ. There is an initial transient associated with the choice of initial point, but because almost all points eventually end up in the attracting set, the Lyapunov exponent is the same for almost all points at a given value of μ. The values of μ for which λ becomes negative in Fig. 2.8 correspond to the regions of periodic behavior evident in Fig. 2.7.

With the numerical results both to stimulate and amaze us, let us try to understand the details of the preceding figures. We first turn our attention to the points of bifurcation. As noted earlier, the first bifurcation takes place at $\mu = 1$. At this value the fixed point at $x = 0$ becomes unstable, and the fixed point at $x_* = (\mu - 1)/\mu$ becomes stable. Although the value of x_* increases with μ through the interval $0 < x_* \leq \frac{2}{3}$ as μ goes from 1 to 3, the qualitative behavior remains the same: x_* is an attracting fixed point for all initial points on the interval (0,1). At $\mu = 3$ the fixed point x_* becomes unstable and an

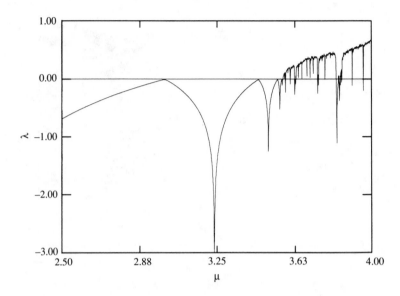

FIGURE 2.8   Lyapunov exponent for the logistic map as a function of the map parameter μ.

attracting 2-cycle is born. In addition to 0 and x_* we find the 2-cycle points x_1, x_2 as the solutions to the equations

$$
\begin{aligned}
x_2 &= \mu x_1 (1 - x_1) \\
x_1 &= \mu x_2 (1 - x_2)
\end{aligned}
$$
(2.16)

resulting in

$$x_{1,2} = (1 + \mu \pm \sqrt{\mu^2 - 2\mu - 3})/2\mu.$$
(2.17)

Note that while the points x_1, x_2 form a 2-cycle for the map F_μ, each is a fixed point for the iterated map F_μ^2.

The stability criterion for a general map in terms of its derivative applies to an iterated map as well. Let the points $x_1, x_2, \ldots, x_p$ denote the points of a p-cycle of an arbitrary one-dimensional map $f\colon [0,1] \to [0,1]$, where the points have been ordered such that $f(x_i) = x_{i+1}$. Each of the points x_i is a fixed point of $f^p(x)$. By the chain rule of differentiation

$$\frac{df^p}{dx}\bigg|_{x_i} = f'(x_1)f'(x_2)\cdots f'(x_p) = \prod_{j=1}^{p} f'(x_j), \quad \text{for all } i = 1, \ldots, p,$$
(2.18)

and thus

$$\left| \frac{df^p}{dx}\bigg|_{x_i} \right| = \prod_{j=1}^{p} |f'(x_j)|, \quad \text{for all } i = 1, \ldots, p.$$
(2.19)

Using this result in the stability condition (2.8) for F_μ^2 gives

$$-1 < \mu^2(1 - 2x_1)(1 - 2x_2) < 1.$$
(2.20)

Substituting from (2.17) for x_1 and x_2 gives $3 < \mu < 1 + \sqrt{6} = 3.44949$ with $dF_\mu^2/dx|_{x_i} = +1$ when $\mu = 3$ and $dF_\mu^2/dx|_{x_i} = -1$ when $\mu = 1 + \sqrt{6}$. When $\mu = 1 + \sqrt{6}$, again we have a loss of stability and bifurcation to a 4-cycle.

An infinite sequence of bifurcations is evident in Fig. 2.7. It is also evident in this figure that the values of μ where these bifurcations take place become ever closer to each other. It becomes also ever more tedious to compute them analytically but Table 2.1 shows the first eight values.

Table 2.1: First 8 Bifurcation Values for μ	
$\mu_1 = 3.0$	$\mu_2 = 3.449490\dots$
$\mu_3 = 3.544090\dots$	$\mu_4 = 3.564407\dots$
$\mu_5 = 3.568759\dots$	$\mu_6 = 3.569692\dots$
$\mu_7 = 3.569891\dots$	$\mu_8 = 3.569934\dots$

Upon examination of this sequence of numbers one is immediately struck by the apparent convergence. Indeed, the convergence is rapid and appears to be geometric. We denote the converged value by μ_∞. Assuming geometric convergence, we write the difference between μ_∞ and μ_k as

$$\mu_\infty - \mu_k = c/\delta^k, \tag{2.21}$$

where c is constant and δ is a constant > 1. A little algebra gives

$$\delta = \frac{\mu_k - \mu_{k-1}}{\mu_{k+1} - \mu_k}.$$

Using the values of μ in Table 2.1, we obtain a rough estimate for $\delta = 4.6692016091\dots$, which is referred to as the Feigenbaum constant after its discoverer (Feigenbaum, 1980a). From (2.21) we can also solve for the constant c and consequently for μ_∞. We find

$$\delta = 4.669202\dots, \quad c = 2.637\dots, \quad \mu_\infty = 3.5699456\dots$$

The interval between μ_∞ and 4 in Fig. 2.7 clearly contains a large number of small windows where the attracting set is a stable periodic cycle. The biggest window corresponds to a stable period 3-cycle. It appears at about $\mu = 3.828427\dots$ and is stable for a significant interval of values in μ. Outside these windows the map looks chaotic.

We can learn much about chaotic maps by looking closely at a very chaotic case. We examine the map for $\mu = 4$ and note for this value of μ that the logistic map is onto for the interval [0,1], and for $\mu > 4$ the map sends points outside this interval of interest.

$$x_{n+1} = 4x_n(1 - x_n). \tag{2.22}$$

Let $x_n = \sin^2 \theta_n$ and substitute into (2.22). After some straightforward algebra we find $\sin^2 \theta_{n+1} = (\sin 2\theta_n)^2$ so that $\theta_{n+1} = 2\theta_n$ and $\theta_n = 2^n \theta_0$. Since $x_0 \in [0,1]$, we may write $\theta_0 = \beta\pi$ where $\beta \in [0,1]$. Furthermore, $\theta_n = \pi 2^n \beta$, and thus it is convenient to write β in binary form:

$$\beta = \sum_{\nu=1}^{\infty} \frac{b_\nu}{2^\nu}, \qquad (2.23)$$

where b_ν equals 1 or 0. The binary representation of β is then given by the infinite string $(b_1, b_2, \ldots)$. With this representation for β we see that the first application of the map gives

$$\theta_0 \to \theta_1 = 2\pi\beta = \pi \sum_{\nu=1}^{\infty} \frac{b_\nu}{2^{\nu-1}} = b_1\pi + \pi \sum_{\nu=1}^{\infty} \frac{b_{\nu+1}}{2^\nu}. \qquad (2.24)$$

Since the sine function is squared, the angles are always modulo π, and so $b_1\pi$ can always be discarded regardless of the value of b_1. Furthermore, since 1 corresponds to the digit string $(11\bar{1}\ldots)$ (where an overbar is placed over a substring that is to be repeated an infinite number of times), the digit strings are always modulo the digit string $(11\bar{1}\ldots)$. This reflects the fact that the angle is always modulo π. Consequently, the map (2.22) or $\theta_n \mapsto \theta_{n+1}$ can be represented by the map $(b_1 b_2 b_3 \ldots) \mapsto (b_2 b_3 b_4 \ldots)$, modulo $(11\bar{1}\ldots)$, i.e., a shift of all binary digits to the left by one slot. Since two nearby initial points will have binary strings that differ in an arbitrary manner, beyond a finite number of slots, all information about the "nearness" of the points will be lost after a finite number of applications of the map. In other words, if the initial point is only specified with finite accuracy, then the digits in its binary representation, after the first few, are random. This means that after a few applications of the map $(b_1 b_2 b_3 \ldots) \mapsto (b_2 b_3 b_4 \ldots)$, all that remains are the random digits! The subsequent applications of the map result in random numbers. Surely this is chaotic dynamics in its most transparent form.

The irrational numbers, and thus almost all x_0, are represented by a string of 0's and 1's that appear as random as the tosses of a coin and are thus aperiodic points. But the rationals are periodic. Using the fact that the digit strings are always modulo the string $(11\bar{1}\ldots)$, we see that $\beta = \frac{1}{3}$, represented by the digit string $(0101\overline{01}\ldots)$ and corresponding to $x = \frac{3}{4}$, is a fixed point. Under the map $(0101\overline{01}\ldots) \mapsto (1010\overline{10}\ldots)$, where $(1010\overline{10}\ldots)$ corresponds to $2\pi/3$, which modulo π corresponds to $\pi/3$. Similarly $\beta = \frac{1}{5}$ is a 2-cycle. These points are unstable in the mapping (2.22), as are all periodic points.

For some values of μ we can calculate, for almost all points, the Lyapunov characteristic exponent. Consider the values of μ for which the single fixed point x_* is the attracting set. If we let the initial point be x_0 and $f^i(x_0) = x_i$, then according to (2.13) we obtain the characteristic exponent λ. But if x_* is the attracting set, then after some finite number of iterations of the map, say M, $x_i \to x_*$. Then we have

$$\lambda(x_0) = \lim_{n \to \infty} \frac{1}{n} \left\{ \sum_{i=0}^{M} \ln|f'(x_i)| + (n - M - 1)\ln|f'(x_*)| \right\} \qquad (2.25)$$

$$= \ln|f'(x_*)| = \ln|2 - \mu|.$$

We see that $\lambda \to 0$ as $\mu \to 1$ or $\mu \to 3$. Furthermore, $\lambda \to -\infty$ as $\mu \to 2$. In a similar way (cf. Exercise 2.5) we see for the range of μ specified by $3 < \mu < 1 + \sqrt{6}$ that $\lambda = \frac{1}{2}\ln|-\mu^2 + 2\mu + 4|$ for the 2-cycle. In this case $\lambda \to 0$ at the ends of the interval, i.e., at $\mu = 3$ and $\mu = 1 + \sqrt{6}$, and $\lambda \to -\infty$ at $\mu = 1 + \sqrt{5}$. This behavior is clearly evident in Fig. 2.8.

2.4 Asymptotic Sets and Bifurcations

With this introduction to the complexities of one-dimensional, noninvertible maps like the tent map (2.2) and the logistic map (2.3), it should be abundantly clear that the dynamics of simple maps is not necessarily simple. In order to analyze this complexity, it is helpful to name and classify important limit sets determined by the maps and to catagorize the possible bifurcations.

We begin by formalizing the notion of an orbit. Consider a map $f: \mathbb{R}^m \to \mathbb{R}^m$ and denote its n-fold composition in the usual way by f^n. Then the *orbit* of $x \in \mathbb{R}^m$ is defined to be

$$\mathcal{O}(x) = \left\{ f^n(x) \right\}_{n=0}^{\infty}. \qquad (2.26)$$

Now we define two special limit points. A point p is said to be an ω-*limit point* (the end) of x if there exist points in $\mathcal{O}(x)$, $f^{n_1}(x), f^{n_2}(x), \ldots$, such that $f^{n_i}(x) \to p$ as $n_i \to \infty$. A point q is said to be an α-*limit point* (the beginning) of x if there exists a sequence of points $x_{n_1}, x_{n_2}, \ldots$, where $f^{n_i}(x_{n_i}) = x$, such that $x_{n_i} \to q$ as $n_i \to \infty$. The α- and ω-limit sets of x, $\alpha(x)$ and $\omega(x)$, are the sets of α- and ω-limit points. Clearly these α- and ω-limit sets belong not only to a specific point but to a specific map. These points represent a beginning (α) and the end (ω) of a point under the given map.

As an example of these limit sets, let us consider the logistic map for $3 < \mu < 1 + \sqrt{6}$ where a stable 2-cycle is the attracting set. Consider the point x_1 of (2.17). The orbit of x_1 is $\mathcal{O}(x_1) = \{x_1, x_2\}$, a set consisting of only two points. The point x_1 and orbit $\mathcal{O}(x_1)$ have period 2. The orbit $\mathcal{O}(x_1)$ is the ω-limit set for all points in $(0,1)$. The α-limit set of x_1 also consists of the points $\mathcal{O}(x_1)$.

We now define a closed set A to be an *attracting set* if it has some neighborhood U such that $f^n(x) \to A$ as $n \to \infty$ for almost all $x \in U$. The union of all such neighborhoods U is called the *basin of attraction* for A. A point

209119

x and its orbit $\mathcal{O}(x)$ are said to be *periodic* of period n, if $f^n(x) = x$ but $f^j(x) \neq x$ for $0 < j < n$. A periodic orbit $\mathcal{O}(x)$ is said to be *stable* if for any neighborhood U of $\mathcal{O}(x)$ there is another neighborhood V so that $f^n(V) \subset U$ for all $n \geq 1$. Of particular interest to us are stable periodic orbits that are also attracting sets. If a periodic orbit is not stable it is said to be *unstable*. An *invariant set* S is define to be a set such that $f(x) \in S$ for all $x \in S$. Perhaps the most basic set for dynamical behavior is the *nonwandering set* Ω. A point x is called *nonwandering* for the map f if given a neighborhood U of x there exists arbitrarily large $n > 0$ such that $f^n(U) \cap U \neq \emptyset$. The set of all nonwandering points constitutes the nonwandering set Ω.

For examples consider again the logistic map and the 2-cycle $\{x_1, x_2\}$ with μ in the range $3 < \mu < 1 + \sqrt{6}$. We now show that the set $\mathcal{O}(x_1)$ is attracting and it is a stable periodic orbit. Both the points x_1 and x_2 are nonwandering and the nonwandering set $\Omega = \mathcal{O}(x_1)$. Let $\delta \ll 1$ and consider the point $x_1 + \delta$ in a small neighborhood about the point x_1.

$$F_\mu(x_1 + \delta) = F_\mu(x_1) + F_\mu'(x_1)\delta + \ldots = x_2 + F_\mu'(x_1)\delta + \ldots.$$

Apply the map F_μ again:

$$F_\mu^2(x_1 + \delta) = x_1 + F_\mu'(x_2)F_\mu'(x_1)\delta + \ldots.$$

The distance away from x_1 is now $|F_\mu'(x_1)F_\mu'(x_2)|\delta = |(F_\mu^2)'(x_1)|\delta$. As long as $|(F_\mu^2)'(x_1)| < 1$, the orbit $\mathcal{O}(x_1)$ is stable and is an attracting set. This condition is the same as we found for the stability of x_1 as a fixed point of F_μ^2. For a periodic orbit $\mathcal{O}(x)$ of period n to be stable the composite map f^n must satisfy (2.8) for all points on the orbit.

Having carefully defined certain useful asymptotic sets, we now turn our attention to the bifurcations of one-dimensional maps. We take the maps in the form $f: \mathbb{R} \times \mathbb{R} \to \mathbb{R}$, where $(\mu, x) \mapsto f_\mu(x) \in \mathbb{R}$. The variable μ is considered the map parameter and x the argument of the map. We use the customary notation C^r to denote the differentiability class of a function. A C^r function is one with its derivatives up to order r being continuous. For brevity we always consider that a coordinate transformation has been done so that the fixed point is at $x = 0$ and that the value of μ for which the fixed point becomes unstable leading to the bifurcation is $\mu = 0$. As discussed previously, the stability is determined by the value of $\partial f/\partial x$, where f is considered a function of two variables, μ and x. For brevity we let $\chi(\mu, x) = \partial f/\partial x$ and let $\chi(0) \equiv \chi(0,0)$. The quantity $\chi(0)$ is just exactly the quantity inside the absolute value signs in (2.8).

The following bifurcation theorems cover the most commonly occuring cases. The names in parentheses are the names frequently given to these bifurcations with sometimes several alternatives. We sketch the proof only for the first theorem so that the reader may see the type of arguments involved. Proofs for the other theorems can be done in a similar way and can be found in Whitley (1986) and Iooss (1979).

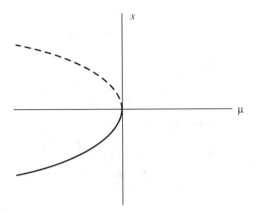

FIGURE 2.9   Curve of fixed points for a fold bifurcation. Unstable points are denoted with a dashed curve.

BF1 (fold, saddle-node, tangent): *Let* $f\colon \mathbb{R} \times \mathbb{R} \to \mathbb{R}$ *be a one-parameter family of* C^2 *maps satisfying:*

(1) $f(0,0) = 0$;   **(2)** $\chi(0) = 1$;   **(3)** $\dfrac{\partial^2 f}{\partial x^2}(0,0) > 0$;   **(4)** $\dfrac{\partial f}{\partial \mu}(0,0) > 0$.

Then there exist intervals $(\mu_1, 0)$, $(0, \mu_2)$ *and* $\epsilon > 0$ *such that*
   (i) if $\mu \in (\mu_1, 0)$, *then* $f_\mu(x)$ *has two fixed points in* $(-\epsilon, \epsilon)$, *with the positive one being unstable and the negative one stable, and*
   (ii) if $\mu \in (0, \mu_2)$, *then* $f_\mu(x)$ *has no fixed points in* $(-\epsilon, \epsilon)$.

See Fig. 2.9 for a sketch of the curve of fixed points in the neighborhood of the origin. Unstable fixed points are noted with a dashed curve. Note that reversing one of the inequalities in hypothesis 3 or 4 reverses the roles of the intervals $(\mu_1, 0)$ and $(0, \mu_2)$.

Proof: Let us define a new function $g(x, \mu) = f(x, \mu) - x$. From the properties of f we find that g satisfies:

$$g(0,0) = 0, \qquad \frac{\partial g}{\partial x}(0,0) = 0, \qquad \frac{\partial^2 g}{\partial x^2}(0,0) > 0, \qquad \frac{\partial g}{\partial \mu}(0,0) > 0.$$

From the classical implicit function theorem [given, for example, in Smith (1983) or in a more elementary form with examples in Devaney (1987, p. 11)] the first and the last of these properties of $g(x, \mu)$ imply the existence of a unique C^2 function $\mu(x)$ for x near 0, with $\mu(0) = 0$ and $g(x, \mu(x)) = 0$. Differentiating this equation we obtain

$$\frac{\partial g}{\partial x} + \frac{\partial g}{\partial \mu}\frac{d\mu}{dx} = 0.$$

Evaluating this at $x = 0$ and using the inequality for $\partial g/\partial \mu$, $d\mu/dx(0) = 0$. Differentiating a second time and evaluating at $x = 0$ results in

$$\frac{\partial^2 g}{\partial x^2}\Big|_{x=0} + \frac{\partial g}{\partial \mu}\Big|_{x=0} \frac{d^2 \mu}{dx^2}\Big|_{x=0} = 0.$$

Coupled with the previous inequalities on $\partial^2 g/\partial x^2$ and $\partial g/\partial \mu$, we conclude $d^2\mu/dx^2(0) < 0$. These results lead to the conclusion that the curve $\mu(x)$ has a maximum at $x = 0$. The only thing remaining is the stability of the branches. The equation $g(x, \mu(x)) = 0$ corresponds to $f(x, \mu(x)) = x$ and gives implicitly the curve of fixed points. For $x > 0$, $d\mu/dx < 0$ and now differentiate the equation for the fixed-point curve in terms of x:

$$\frac{df}{dx} = \frac{\partial f}{\partial x} + \frac{\partial f}{\partial \mu}\frac{d\mu}{dx} = 1, \quad \text{and} \quad \frac{\partial f}{\partial x} = 1 - \frac{\partial f}{\partial \mu}\frac{d\mu}{dx}.$$

For $x = \epsilon > 0$, $d\mu/dx < 0$ and $\partial f/\partial \mu > 0$, which implies that $\partial f/\partial x$ at $x = \epsilon$ is greater than 1 and hence unstable. A similar argument shows that the negative fixed point is stable.

BF2 (flip): Let $f: \mathbb{R} \times \mathbb{R} \to \mathbb{R}$ be a one-parameter family of C^3 maps satisfying

$$(1) \ f(0,0) = 0, \quad (2) \ \chi(0) = -1.$$

Then there exists a unique branch of fixed points $x(\mu)$ for μ near 0 with $x(0) = 0$. Furthermore, if $\chi(\mu) = \partial f/\partial x(x(\mu), \mu)$ satisfies

$$(3) \ \frac{d\chi}{d\mu}(0) < 0 \quad \text{and} \quad (4) \ \frac{\partial^3 f^2}{\partial x^3}(0,0) < 0.$$

Then there are intervals $(\mu_1, 0)$, $(0, \mu_2)$, and $\epsilon > 0$ such that
   (i) if $\mu \in (0, \mu_2)$, then $f_\mu(x)$ has one unstable fixed point and one stable orbit of period 2 for $x \in (-\epsilon, \epsilon)$, and
   (ii) if $\mu \in (\mu_1, 0)$, then $f_\mu(x)$ has a single stable fixed point for $x \in (-\epsilon, \epsilon)$.

See Fig. 2.10 for a sketch of this case.
Remarks
   1. Changing the direction of the inequality on hypothesis 3 changes the stability of the fixed point to being first unstable and then passing through 0 to stability.
   2. Changing the sense of the inequality in hypothesis 4 makes all the period 2 points unstable rather than stable. This is sometimes referred to as a subcritical bifurcation and is illustrated in Fig. 2.11. Note that the period-doubling bifurcations of the logistic map are flip bifurcations.

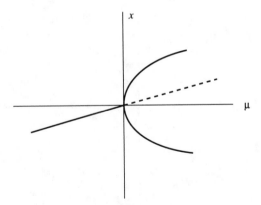

FIGURE 2.10   Curves of fixed points for a flip bifurcation.

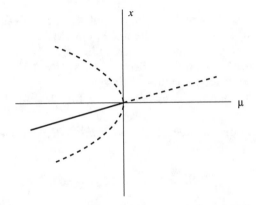

FIGURE 2.11   Bifurcation curves for a subcritical flip bifurcation.

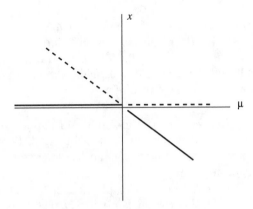

FIGURE 2.12   Transcritical bifurcation curve.

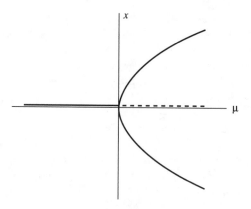

FIGURE 2.13   Pitchfork bifurcation diagram.

These theorems more or less take care of the general case with the slope becoming +1 (fold) or −1 (flip) at the fixed point. Adding special symmetries or other restrictions on the function gives additional types of bifurcations. We consider two types that commonly occur with $\chi(0) = 1$. In the first case the origin is a fixed point for all values of the parameter μ. This is contained in hypothesis 1 of the following proposition. See Fig. 2.12 for a sketch.

BF3 (transcritical): *Let $f: \mathbb{R} \times \mathbb{R} \to \mathbb{R}$ be a one-parameter family of C^2 maps satisfying:*

$$(1) f(0, \mu) = 0, \qquad (2)\ \chi(0) = 1,$$

$$(3)\ \chi(\mu) = \frac{\partial f}{\partial x}(0, \mu) \quad \text{with} \quad \frac{d\chi}{d\mu}(0) > 0, \qquad (4)\ \frac{\partial^2 f}{\partial x^2}(0,0) > 0.$$

Then f has two branches of fixed points for μ near zero. The first branch is $x = 0$ for all μ, and the second, bifurcated branch has $x(\mu) \neq 0$ if $\mu \neq 0$, but $x(0) = 0$. The first branch with $x = 0$ is stable if $\mu < 0$ and unstable if $\mu > 0$, while the fixed points on the bifurcated branch have the opposite stability.

Note that changing the sense of the inequality in hypothesis 4 changes the slope $dx/d\mu$ of the bifurcated solution branch in Fig. 2.12 from negative to positive. The bifurcation that takes place in the logistic map at $\mu = 2$ is a transcritical bifurcation.

A second kind of restriction on a map is that it be an odd function, i.e., $f(-x, \mu) = -f(x, \mu)$. In this case it is obviously true that $f(0, \mu) = 0$ but BF3 does not apply because $\partial^2 f/\partial x^2$ is also odd implying $\partial^2 f/\partial x^2(0, \mu) = 0$ and hence hypothesis 4 of BF3 is not met. Instead we obtain the following proposition.

BF4 (pitchfork): *Let $f: \mathbb{R} \times \mathbb{R} \to \mathbb{R}$ be a one-parameter family of C^3 maps satisfying:*

$$\textbf{(1) } f(-x, \mu) = -f(x, \mu); \quad \textbf{(2) } \chi(0) = 1;$$

$$\textbf{(3) } \chi(\mu) = \frac{\partial f}{\partial x}(0, \mu) \text{ with } \frac{d\chi}{d\mu}(0) > 0; \quad \textbf{(4) } \frac{\partial^3 f}{\partial x^3}(0, 0) < 0.$$

Then there are intervals $(\mu_1, 0)$, $(0, \mu_2)$, and $\epsilon > 0$ such that:
(i) if $\mu \in (\mu_1, 0)$, then $f_\mu(x)$ has a single stable fixed point at the origin in $(-\epsilon, \epsilon)$, and
(ii) if $\mu \in (0, \mu_2)$, then f_μ has three fixed points in $(-\epsilon, \epsilon)$. The origin $x = 0$ is an unstable fixed point while the other two fixed points are stable.

See Fig. 2.13 for a sketch. Note that the bifurcation diagrams for the flip and the pitchfork bifurcations look very similar, but the stable curves for the flip represent a 2-cycle while the stable curves for the pitchfork represent two distinct fixed points.

The reader is again strongly encouraged to use a programmable calculator to verify the asserted behavior of the example maps in Table 2.2. There is simply no substitute for computational experience.

Type	$\frac{\partial f}{\partial x}(0,0)$	Fixed Point	Other Hypotheses	Examples
Fold (tangent, saddle-node)	1	$f(0,0) = 0$	$\frac{\partial^2 f}{\partial x^2}(0,0) > 0, \ \frac{\partial f}{\partial \mu}(0,0) > 0$	$\mu + x + x^2$
Flip	-1	$f(0,0) = 0$	$\frac{d\chi}{d\mu}(0) < 0, \ \frac{\partial^3 f^2}{\partial x^3}(0,0) < 0^a$	$\mu - x - x^2$
Transcritical	1	$f(\mu, 0) = 0$	$\frac{d\chi}{d\mu}(0) > 0, \ \frac{\partial^2 f}{\partial x^2}(0,0) > 0^b$	$(\mu+1)x + x^2$
Pitchfork	1	$f(\mu, -x) = -f(\mu, x)$	$\frac{d\chi}{d\mu}(0) > 0, \ \frac{\partial^3 f}{\partial x^3}(0,0) < 0^b$	$(\mu+1)x - x^3$

Table 2.2: Bifurcation Table for $f(\mu, x)$

^{a} $\chi(\mu) = \partial f/\partial x(\mu, x(\mu))$, and $x(\mu)$ is the curve of fixed points, not period 2 points.
^{b} $\chi(\mu) = \partial f/\partial x(\mu, 0)$

Exercises

2.1 Consider a tent map $\Delta_\mu(x)$ for $\frac{1}{2} < \mu < 1$ (for example, $\mu = 0.75$). Use graphical analysis on the first few iterates and induction (or any other valid way) to show that the linear pieces in $\Delta_\mu^n(x)$ have slope $\pm(2\mu)^n$. Conclude that two points initially separated by a distance ϵ, after n iterations of $\Delta_\mu(x)$ are separated by a distance $\epsilon(2\mu)^n$. Show this to be consistent with a Lyapunov exponent for this map equal to $\ln(2\mu)$. Thus even for values of $\mu < \frac{1}{2}$ the map $\Delta_\mu(x)$ is chaotic even though in the resolution of Fig. 2.3 it looks like there is a period 2 attractor.

2.2 Show by numerical experimentation that the map $f(x) = \mu \sin \pi x$ leads to the same value for the Feigenbaum constant δ.

2.3 Verify that $\beta = \frac{1}{3}$ corresponds to a fixed point by consideration of the binary shift operation. Find the binary representation of $\beta = \frac{1}{5}$, verify that it is a 2-cycle, and find the points on the interval [0,1] corresponding to this cycle.

2.4 With the map parameter $\mu = 1$ show that the tent map is also equivalent to the binary shift map $(b_1 b_2 b_3 \ldots) \mapsto (b_2 b_3 b_4 \ldots)$.

2.5 For μ values where the attracting set consists of the 2-cycle points of (2.17), show that the Lyapunov characteristic exponent λ is given by $\frac{1}{2} \ln | - \mu^2 + 2\mu + 4|$.

2.6 Prove the statement about reversing the inequalities following the statement of BF1. What simple change in the direction of the inequalities is required to make the negative fixed point unstable and the positive one stable?

2.7 Describe the bifurcation of the function $f_\mu(x) = \mu e^x$ for $\mu > 0$ as μ passes through the value $1/e$.

2.8 Describe the bifurcation of the function $f_\mu(x) = \mu e^x$ for $\mu < 0$ as μ passes through the value $-e$.

2.9 Sketch the bifurcation curves for the four example maps in Table 2.2.

2.10 Consider the so-called *Baker's Transformation*

$$x_{n+1} = 2\mu x_n \ (\text{mod} 1)$$

on the interval [0,1]. Sketch the map and investigate its fixed points, their stability, and the Lyapunov exponents for the map.

CHAPTER THREE

UNIVERSALITY THEORY

Such glances! Such impudence!
This flower has the effect
Of a bullet striking me!
Its perfume is strong, and the flower is pretty!
(Carmen, Act I)

In the last chapter we focused on the transistion to chaos observed in some one-dimensional maps as a parameter of the map was varied. In particular we saw that the logistic map produced a sequence of period doublings culminating in chaos. More precisely, we observed a doubling of the number of points in the periodic attractors until the attracting set for almost all points on the interval appeared to be an infinite chaotic set. The behavior of the iterates suggested that the dynamics was chaotic, and the Lyapunov characteristic exponent also indicated chaos. In this chapter we delve deeper into the behavior of the logistic map and its iterates as this path to chaos is traversed. We then conclude the chapter with an introduction to intermittency and the intermittent transistion to chaos.

3.1 Period Doubling and the Composite Functions

There is probably no better way to discover the fundamental mechanism for period doubling than by considering simultaneously the function $F_\mu^{2^n}$ and its doubled counterpart $F_\mu^{2^{n+1}} = F_\mu^{2^n} \circ F_\mu^{2^n}$. We begin with $n = 0$. Figures 3.1a and 3.1b show F_μ and F_μ^2 for a value of μ below the bifurcation value $\mu = 3.0$. It is clear that the curve representing the function values intersects the line $y = x$ at a single point, the fixed point x_*. Furthermore, by laying a straightedge along the curve and comparing with the diagonal lines, it is easy to see that $|dF_\mu/dx(x_*)| < 1$. Now compare Fig. 3.1 with Fig. 3.2 in which μ has been increased to a value above 3.0. Note that $|dF_\mu/dx(x_*)|$ has increased beyond 1 and that two new intersection points of F_μ^2 with the line $y = x$ have developed. These new intersection points for F_μ^2 represent a 2-cycle for F_μ. The point x_* remains a fixed point both for F_μ and F_μ^2 but has become unstable.

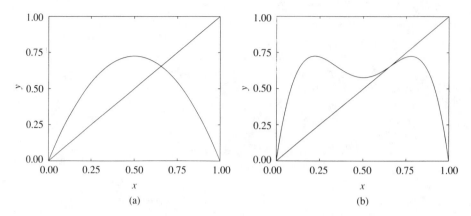

FIGURE 3.1   Function values for (a) F_μ and (b) F_μ^2 for a value of $\mu = 2.90$ below the bifurcation value $\mu = 3.0$.

This is the pattern that repeats itself as μ increases. With increasing μ the 2^n-cycle of F_μ corresponding to a set of fixed points of $F_\mu^{2^n}$ becomes unstable and a new set of intersection points of $F_\mu^{2^{n+1}}$ with $y = x$ develop giving a stable 2^{n+1}-cycle. Figure 3.3 and 3.4 show again this pattern on each side of $\mu = 3.44949\ldots$. At the stable cycle points the derivative increases beyond 1 in absolute magnitude as μ passes through the bifurcation value and a new stable cycle develops with twice the number of points in the cycle.

Between bifurcation values for the map parameter, such as those listed Table 2.1, there exist values of μ for which the 2^n-cycle is said to be *superstable* and it is customary to refer to them as *supercycles*. These supercycles have maximum stability in that $|dF_\mu^{2^n}/dx|$ is as small as possible, namely 0. From (2.19) we know that if $|dF_\mu^{2^n}/dx(x_i)| = 0$, then one of the points on the 2^n-

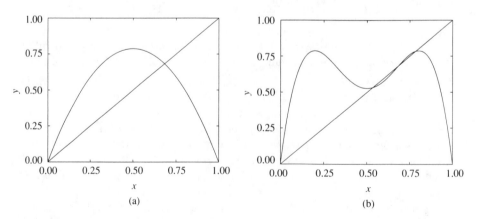

FIGURE 3.2   Function values for $(a)F_\mu$ and $(b)F_\mu^2$ for a value of $\mu = 3.15$ that is above the bifurcation value of $\mu = 3.0$.

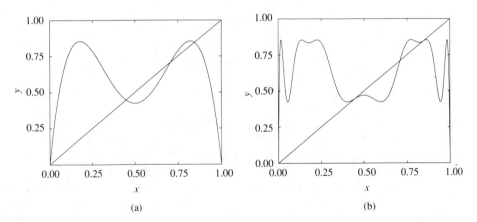

FIGURE 3.3   The function $(a)F_\mu^2$ and $(b)F_\mu^4$ for a value of $\mu = 3.42$ that is below the second bifurcation value $\mu_2 = 3.44949...$.

cycle must be $\bar{x} \equiv \frac{1}{2}$. A 2^n-supercycle is a set of 2^n points in the interval $(0,1)$ forming a 2^n-cycle, and one of these points is $\bar{x}$. The 2^n-supercycle is the orbit of $\bar{x}$, i.e., $\mathcal{O}(\bar{x})$. We label the map parameter corresponding to such a supercycle with an overbar, i.e., $\bar{\mu}$. For $\mu = \bar{\mu}_n$, where

$$\mu_n < \bar{\mu}_n < \mu_{n+1}, \tag{3.1}$$

we have a 2^n-supercycle with

$$\left| \frac{dF_{\bar{\mu}_n}^{2^n}(x_i)}{dx} \right| = \prod_{j=0}^{2^n-1} \left| F_{\bar{\mu}_n}'(x_j) \right| = 0, \tag{3.2}$$

where x_i is any point in the supercycle.

Aside from the point $\bar{x}$ in the supercycle, there is one other point on which we focus our attention. This is the point nearest to $\bar{x}$; we call this point x_1.

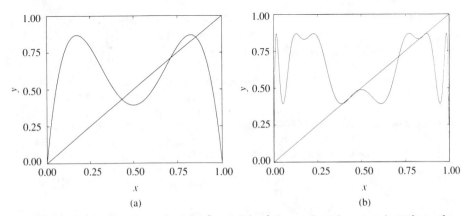

FIGURE 3.4   The functions (a) F_μ^2 and (b) F_μ^4 for a value of $\mu = 3.48$ just above the second bifurcation value $\mu_2 = 3.44949...$.

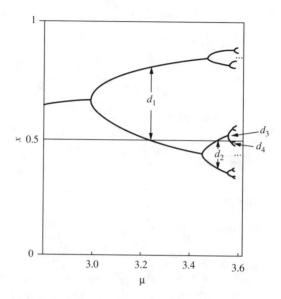

FIGURE 3.5   Bifurcation diagram for the logistic map showing the distance between nearest neighbor cycle elements in the supercycles.

This point x_1 must be the corresponding point that split out of $\bar{x}$ as the map parameter passed through μ_n. For well-defined maps image points must be unique, and so there cannot be any "crossings," as is evident in Fig. 2.7. Just before $\mu = \mu_n$, the point x_0, out of which $\bar{x}$ and x_1 evolved, was a fixed point of $F_\mu^{2^{n-1}}$. But once μ passes through μ_n and increases toward $\bar{\mu}_n$, x_0 is no longer a stable fixed point of $F_\mu^{2^{n-1}}$ but has spawned two elements of a stable 2-cycle for $F_\mu^{2^{n-1}}$. Thus we see that x_1 is the point given by

$$F_{\bar{\mu}_n}^{2^{n-1}}(\bar{x}) = x_1. \tag{3.3}$$

With this characterization of the point x_1 in hand, we examine the distance between $\bar{x}$ and x_1 for increasing values of n. We define

$$d_n = F_{\bar{\mu}_n}^{2^{n-1}}(\bar{x}) - \bar{x}. \tag{3.4}$$

For reasons of brevity it is convenient to shift the origin to $\bar{x} = \frac{1}{2}$, i.e., to the maximum of the logistic map F_μ. We define $\xi = x - \frac{1}{2}$ and simply use the same symbol for the function of ξ as we have used for $F_\mu(x)$. Thus the distance d_n is given by

$$d_n = F_{\bar{\mu}_n}^{2^{n-1}}(0). \tag{3.5}$$

These distances are pictured schematically in Fig. 3.5, which is an adaptation from Fig. 2.7.

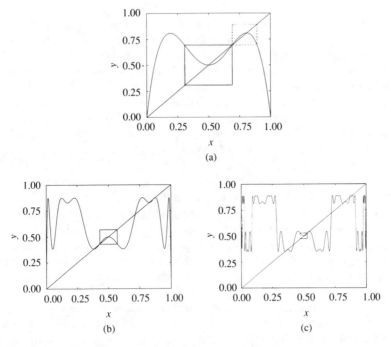

FIGURE 3.6   Iterates of the logistic map for values of μ corresponding to supercycles:
(a) The 2^1 iterate with $\mu = 3.24...$, (b) the 2^2 iterate with $\mu = 3.50...$, and (c) the 2^3 iterate
with $\mu = 3.55...$. The little boxes in the center enclose the maps similar to the logistic map
when properly rescaled.

We note in Fig. 3.5 that the distances d_n alternate in being first above the
line $x = \frac{1}{2}$ and then below this line. Furthermore, a numerical computation
of these distances shows that their ratio converges toward a constant:

$$\frac{d_n}{d_{n+1}} \simeq -\alpha = -2.50280... \qquad \text{for large } n. \tag{3.6}$$

The minus sign in (3.6) takes account of the fact that d_n as defined in (3.5) is
first positive and then negative, then positive, etc. From (3.6) it follows that
$d_1 \simeq (-\alpha)^n d_{n+1}$ or using (3.5)

$$d_1 = F_{\bar{\mu}_1}(0) \simeq \lim_{n \to \infty} (-\alpha)^n F_{\bar{\mu}_{n+1}}^{2^n}(0). \tag{3.7}$$

This result is extremely suggestive because it says that if we scale the
peak value of $F_{\bar{\mu}_{n+1}}^{2^n}$ by an appropriate amount for each n, then these values
converge to a limit.

There is one other important observation to make about the flip bifurcation sequence. Each successive iterate of F_μ becomes unstable in exactly
the same way as its predecessor. That is to say, $F_\mu^{2^n}$ becomes unstable at
$\mu = \mu_{n+1}$ in exactly the same way that $F_\mu^{2^{n-1}}$ became unstable at $\mu = \mu_n$.

This is particularly evident when one focuses attention on the cycle point nearest $\bar{x} = \frac{1}{2}$. Figure 3.6 shows a series of iterates of F_μ at values of μ corresponding to supercycles. The width of the little boxes is given by the distance from $\bar{x}$ to the next nearest fixed point. If we invert the boxes for odd n, we see that inside the boxes the curve of function values looks very much like the original logistic map. Since all elements of the supercyle are slaved to $\bar{x}$ by $F_{\bar{\mu}_{n+1}}^{2^n}$, all relative distances should scale in the same way. Thus these distances beteen $\bar{x}$ and the nearest fixed point also scale by the factor α.

Not only is this true for the function iterates depicted in Fig. 3.6 but for subsequent supercycles, as can be seen if one magnifies the appropriate part of the curve of function values near $\bar{x} = \frac{1}{2}$. This result suggests that these functions are converging toward a *universal* function.

To get some idea how this convergence toward a universal function works, we consider a renormalization calculation that does not make use of the supercycle information, but rather only uses the information that there is an infinite sequence of flip bifurcations. On the first curve of Fig. 3.6 a box enclosed with dashed lines suggests the following renormalization procedure for the logistic map.

1. Rescale the map so that the flip bifurcation takes place at $x = 0$, $\mu = 0$. For the logistic map this means moving the origin to x_* and using a new map parameter that is 0 when $\mu = 3$. Using the new coordinate z and map parameter ν, the logisitic map becomes

$$f_\nu(z) = -(\nu + 1)z - (\nu + 3)z^2, \qquad (3.8)$$

   where $z = x - x_*$ and $\nu = \mu - 3$. It is straightforward to check that $f_\nu(z)$ in (3.8) satisfies the hypothesis of BF2 for a flip bifurcation at $\nu = 0$ and $z = 0$.

2. Find the 2-cycle points for (3.8) that are the fixed points for $f_\nu^2(z)$. The positive fixed point is $z_1 = (\sqrt{\nu^2 + 4\nu} - \nu)/2(\nu + 3)$, corresponding to x_1 of (2.17) in the new coordinates.

3. Shift the origin in z to this fixed point z_1 of f_ν^2, i.e., change to a new coordinate $\tilde{z} = z - z_1$, and adjust the function so that $\tilde{z} = 0$ is a fixed point. The new function is then $\bar{f}_\nu(\tilde{z}) = f_\nu^2 - z_1$, where $z = \tilde{z} + z_1$, and z_1 is given in terms of ν.

4. With some algebra this function $\bar{f}_\nu(\tilde{z})$ can be written in the form

$$\bar{f}_\nu(\tilde{z}) = c_1\tilde{z} + c_2\tilde{z}^2 + \cdots$$

   or

$$\tilde{z}_{n+1} = c_1\tilde{z}_n + c_2\tilde{z}_n^2 + \cdots,$$

   where c_1 and c_2 are functions only of ν.

5. Rescale the independent variable; $\bar{z} = a\tilde{z}$, then

$$\bar{z}_{n+1} = c_1\bar{z}_n + \frac{c_2}{a}\bar{z}_n^2 + \cdots.$$

Now demand that this equation be exactly of the same form as (3.8) in order to get $\bar{\nu}$, the new map parameter. This gives

$$-(\bar{\nu}+1) = c_1 \quad \text{and} \quad -(\bar{\nu}+3) = \frac{c_2}{a}.$$

These equations can be solved for $\bar{\nu}$ and a in terms of ν giving

$$\bar{\nu} = \nu^2 + 4\nu - 2 \tag{3.9}$$

and

$$a = \frac{\nu^3 + 7\nu^2 + 3\sqrt{\nu^2 + 4\nu}(\nu + 3) + 12\nu}{\nu^2 + 4\nu + 1}. \tag{3.10}$$

Thus having the map in exactly the same form as (3.8), with new coordinate $\bar{z}$ and new map parameter $\bar{\nu}$, we can repeat this entire process, renormalizing at each stage. Eventually this process converges to a specific value for the map parameter ν_∞ and a rescaling parameter a_∞.

We can view (3.9) as giving an estimate for the equation that ν_∞ must satisfy.

$$\nu_\infty = \nu_\infty^2 + 4\nu_\infty - 2.$$

Solving for ν_∞ gives $\nu_\infty = 0.5615528\ldots$. Adding the original shift value of 3 to this we obtain $\mu_\infty = 3.5615528\cdots$, a result remarkably close to the more exact value given following Eq. (2.21). Also $d\bar{\nu}/d\nu$ from (3.9) evaluated at ν_∞ gives an approximation to the Feigenbaum constant δ [cf. equation following (2.21)]. The constant a evaluated at ν_∞ gives an analytic estimate for the Feigenbaum constant α for scaling distances. Renormalization of the map given in Exercise 3.1 gives at this first stage of renormalization a better estimate of α.

This simple and straightforward renormalization of a bifurcating map illustrates the renormalization process and the constant rescaling leading to the Feigenbaum constants. The scaling comes about through functional composition and suggests the existence of some function that when composed with itself will reproduce itself, when properly scaled.

Returning now to the supercycles, the solid boxes in Fig. 3.6 suggest the following renormalization scheme for this function with a quadratic maximum at $\xi = 0$: reduce the scale in ξ by a factor $(-1/\alpha)$ and multiply the values of the function composed with itself by $-\alpha$ at each stage in order to make the curves of function values look the same. At each stage of this process α will have a particular value. In the limit we expect this process to lead to a function that upon renormalization gives itself back again. This will happen for special α. That is

$$g(\xi) = (-\alpha)g(g(\xi/-\alpha)). \tag{3.11}$$

Since the function $g(\xi)$ is quadratic, we could discard the minus sign in the argument of the inner g. We also note that the actual scale of this universal function is arbitrary. This follows because if $g(\xi)$ satisfies (3.11), then so does $\bar{g}(\xi) \equiv ag(\xi/a)$, with a an arbitrary scale factor. Thus at $\xi = 0$ we can fix the scale arbitrarily. We take

$$g(0) = 1 \qquad (3.12)$$

and find

$$g(1) = -1/\alpha. \qquad (3.13)$$

The function $g(\xi)$ is a particular function of ξ^2 satisfying (3.11) along with (3.12) and (3.13). We can obtain a polynomial approximation to this function by taking

$$g(\xi) = 1 + a_1\xi^2 + a_2\xi^4 + a_3\xi^6 + \cdots. \qquad (3.14)$$

Then substitute into (3.11) and match the coefficients for all powers of ξ. This gives values for the coefficients and α. This is a tedious process if one proceeds analytically beyond the first one or two coefficients, but Feigenbaum (1979) has given the following list of numerically determined coefficients accurate to ten significant figures.

$$\alpha = 2.502907876 \qquad a_4 = -3.527413864 \times 10^{-3}$$
$$a_1 = -1.527632997 \qquad a_5 = 8.158191343 \times 10^{-5}$$
$$a_2 = 1.1048151943 \times 10^{-1} \qquad a_6 = 2.536842339 \times 10^{-5}$$
$$a_3 = 2.670567349 \times 10^{-2} \qquad a_7 = -2.687772769 \times 10^{-6}$$

The universal function $g(\xi)$ is a "fixed point" in the space of functions for this operation of function composition followed by rescaling. We formalize this operation by defining the doubling transformation T. The proper definition for this operator on the space of functions is obtained by considering the limit of the supercycle functions. Taking the limit as $n \to \infty$ of the process represented by the series of solid boxes in Fig. 3.6, we define

$$g_1(\xi) = \lim_{n \to \infty} (-\alpha)^n F_{\bar{\mu}_{n+1}}^{2^n}(\xi/(-\alpha)^n). \qquad (3.15)$$

We note that (3.15) is just the generalization of (3.7) to the entire interval. We can in fact further generalize this relation in an important way by replacing the subscript 1 in (3.15) with i. That is, we define

$$g_i(\xi) = \lim_{n \to \infty} (-\alpha)^n F_{\bar{\mu}_{n+i}}^{2^n}(\xi/(-\alpha)^n). \qquad (3.16)$$

Then examine the function $g_{i-1}(\xi)$.

$$g_{i-1}(\xi) = \lim_{n \to \infty} (-\alpha)^n F_{\bar{\mu}_{n+i-1}}^{2^n}\left(\xi/(-\alpha)^n\right)$$
$$= \lim_{n \to \infty} (-\alpha)(-\alpha)^{n-1} F_{\bar{\mu}_{n-1+i}}^{2^{n-1+1}}\left(\frac{-1}{\alpha}\left(\xi/(-\alpha)^{n-1}\right)\right). \qquad (3.17)$$

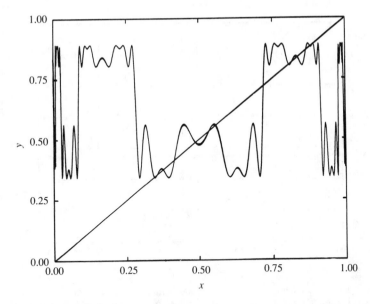

FIGURE 3.7   Superposition of high-order approximations to the universal functions g_1, g_2, and g_3. The deepest extrema correspond to the highest-order function.

Now replace $n - 1$ by m and note that $F^{2^{m+1}} = F^{2^m} \circ F^{2^m}$. Thus (3.17) becomes

$$g_{i-1}(\xi) = \lim_{n \to \infty} (-\alpha)(-\alpha)^m F_{\bar{\mu}_{m+i}}^{2^m}\left(\frac{1}{(-\alpha)^m}(-\alpha)^m F_{\bar{\mu}_{m+i}}^{2^m}\left(\frac{-1}{\alpha}(\xi/(-\alpha)^m)\right)\right). \tag{3.18}$$

Referring to (3.16), we see that this can be written in the form

$$g_{i-1}(\xi) = (-\alpha)g_i(g_i(\xi/-\alpha)). \tag{3.19}$$

This suggests the definition for the doubling transformation:

$$T[g_i(\xi)] = (-\alpha)g_i^2(\xi/-\alpha) = (-\alpha)g_i(g_i(\xi/-\alpha)) = g_{i-1}(\xi). \tag{3.20}$$

As an example

$$g_0(\xi) = T[g_1(\xi)] = (-\alpha)g_1(g_1(\xi/-\alpha)). \tag{3.21}$$

Thus we have a whole family of universal functions $g_i(\xi), i = 0, 1, \ldots$.

How do these functions differ? We can ascertain the answer by looking at high-order approximations to g_1, g_2, and g_3. We first note that Fig. 3.6c represents the first approximation and indicates the size of the renormalization box. Figure 3.7 superimposes the curves for g_1, g_2, and g_3 to the same level of approximation. Although just barely perceptible in the given resolution, the curves have successively deeper extrema. Referring to Fig. 3.8, we see

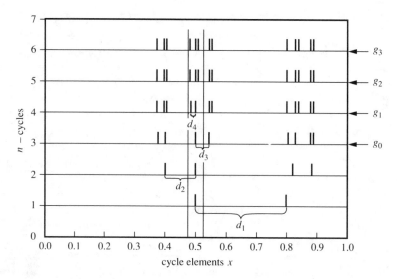

FIGURE 3.8   Location of the cycle elements in successively higher-order supercycles. The vertical lines delineate the approximate size of the renormalization box for the 2^3-order approximation to the universal functions. The levels corresponding to the universal functions are indicated on the left.

that the function g_0 locates the cycle elements with one cycle element at the extrema, g_1 locates the same cycle elements by determining two elements per extremum, and g_2 locates again the same elements by locating $2^2 = 4$ elements per extremum. Continuing in this fashion $g_i(\xi)$ locates 2^i elements about each extremum near a fixed point. Figure 3.8 indicates the cycle elements in the renormalization box of the 2^3 approximation to the universal functions.

Each F^{2^n} is always magnified by the same $(-\alpha)^n$ for every i, and thus the scales of the g_i in Fig. 3.7 are the same. Indeed, each g_i looks essentially like g_1, except the peaks are successively a little sharper in order to produce a larger number of elements of the cycle at each extremum. The amount by which each of these extrema must grow to accomodate a larger number of elements in the cycle becomes ever smaller. We conclude that the limit

$$\lim_{i \to \infty} g_i(\xi) = g(\xi) \tag{3.22}$$

exists. That this limit is exactly $g(\xi)$ of (3.11) follows immediately by taking the limit of (3.19).

The function $g(\xi)$ is a universal function locating *all* cycle elements near its central maximum. Furthermore, from (3.22) and (3.16) we have

$$g(\xi) = \lim_{n \to \infty} (-\alpha)^n F_{\mu_\infty}^{2^n}(\xi/(-\alpha)^n). \tag{3.23}$$

In contrast to the $g_i(\xi)$, $g(\xi)$ is obtained as a limit of logistic map iterates at a *fixed* value of μ. This provides another characterization of the parameter

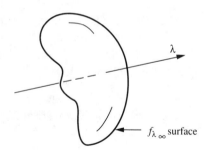

FIGURE 3.9   Sketch to suggest the stable and unstable manifolds of the fixed point g.

μ_∞. It is the special value of μ for which the limit of functional iteration and magnification lead to a convergent function.

The function $g(\xi)$ is a fixed point in function space of the doubling operator T, which acts on functions according to (3.20). That is to say

$$T[g] = g. \tag{3.24}$$

However, (3.20) shows that this is also an *unstable* fixed point in the space of functions. If g_i for large i is a close approximation to g in some function norm, then T applied to g_i gives $g_{i-1} = T[g_i]$, which is further from g than g_i was. The unstable nature of this fixed point can also be exploited to calculate the universal functions g_i. We assume that we have a knowledge of $g(\xi)$ by some approximate solution of (3.11), such as in (3.14). We then can find g_i for large i by linearizing T about its fixed point g. Then repeated applications of T to g_i provides g_0 or g_1.

How unstable is the fixed point g of the doubling operator T? We consider a qualitative argument due to Feigenbaum (1980a) and think of a one-parameter family of functions f_λ, which undergo the period-doubling sequence of bifurcations. Two examples of such families on the interval $[0,1]$ are the logistic map of (2.14) and the sine map of Exercise 2.2. Such a one-parameter family corresponds to a "line" in function space. For each family (line) there is an isolated parameter value λ_∞ for which repeated applications of T lead f_{λ_∞} to converge toward g. For a sketch suggesting these ideas see Fig. 3.9.

As an example we see from (3.20) that

$$T^n\left[F_{\mu_\infty}\right](\xi) = (-\alpha)^n F_{\mu_\infty}^{2^n}\left(\xi/(-\alpha)^n\right),$$

and thus from (3.23)

$$\lim_{n \to \infty} T^n\left[F_{\mu_\infty}\right](\xi) = g(\xi). \tag{3.25}$$

The function space can be "packed" with all the lines corresponding to various function families. The set of all points on these lines specified by the respective λ_∞'s determines a "surface" in function space having the property that repeated application of T to any function on this surface coverges toward g. This surface is the stable manifold of T through g. But through each point of this surface issues out the corresponding line, which is one-dimensional since it is parameterized by a single parameter λ. Consequently, we conclude that T is unstable near g in only one direction in function space. Numerical explorations and proofs of this issue have been reported by Feigenbaum (1980a), Collet et al. (1980), and Lanford (1982).

Let us now linearize the doubling operator T near g. We consider $\bar{\mu}$ to be some $\bar{\mu}_n$ for $n \gg 1$, i.e., $\bar{\mu}$ is near μ_∞, and write

$$F_{\bar{\mu}}(\xi) \simeq F_{\mu_\infty} + (\mu_\infty - \bar{\mu})DF(\xi), \qquad (3.26)$$

where

$$DF(\xi) = \frac{\partial F_{\bar{\mu}}(\xi)}{\partial \bar{\mu}}\Big|_{\mu_\infty}. \qquad (3.27)$$

Now apply the doubling operator T to (3.26).

$$T[F_{\bar{\mu}}](\xi) = -\alpha F_{\bar{\mu}}\left(F_{\bar{\mu}}\left(\xi/(-\alpha)\right)\right) \qquad (3.28)$$

and

$$F_{\bar{\mu}}(\xi/(-\alpha)) \simeq F_{\mu_\infty}(\xi/(-\alpha)) + (\bar{\mu} - \mu_\infty)DF(\xi/(-\alpha))$$

$$F_{\bar{\mu}}\left(F_{\bar{\mu}}\left(\xi/(-\alpha)\right)\right) \simeq F_{\mu_\infty}\left(F_{\mu_\infty}\left(\xi/(-\alpha)\right) + (\bar{\mu} - \mu_\infty)DF\left(\xi/(-\alpha)\right)\right)$$

$$+ (\bar{\mu} - \mu_\infty)DF\left(F_{\mu_\infty}\left(\xi/(-\alpha)\right)\right.$$

$$\left. + (\bar{\mu} - \mu_\infty)DF\left(\xi/(-\alpha)\right)\right)$$

$$\simeq F_{\mu_\infty}\left(F_{\mu_\infty}\left(\xi/(-\alpha)\right)\right)$$

$$+ F'_{\mu_\infty}\left(F_{\mu_\infty}\left(\xi/(-\alpha)\right)\right)(\bar{\mu} - \mu_\infty)DF\left(\xi/(-\alpha)\right)$$

$$+ (\bar{\mu} - \mu_\infty)DF\left(F_{\mu_\infty}\left(\xi/(-\alpha)\right)\right)$$

Using (3.28)

$$T[F_{\bar{\mu}}](\xi) \simeq T[F_{\mu_\infty}](\xi) - \alpha(\bar{\mu} - \mu_\infty)\left\{F'_{\mu_\infty}\left(F_{\mu_\infty}(\xi/(-\alpha))\right)DF(\xi/-\alpha)\right.$$

$$\left. + DF\left(F_{\mu_\infty}(\xi/-\alpha)\right)\right\}, \qquad (3.29)$$

where only lowest order terms in $(\bar{\mu} - \mu_\infty)$ have been kept. From (3.29) we can read off the linearized piece of T and how it depends on the function about which it is linearized. We denote this linearized piece as $L\{F_{\mu_\infty}\}[DF](\xi)$, where the notation conveys the fact that L depends on the "point" in function space about which we linearize (the argument in braces), it "operates" on the function in the brackets, and the result is a function on the domain of the variable ξ. From (3.29)

$$L\{F_{\mu_\infty}\}[DF](\xi) = -\alpha\left\{F'_{\mu_\infty}\left(F_{\mu_\infty}\left(\frac{\xi}{-\alpha}\right)\right)DF\left(\frac{\xi}{-\alpha}\right) + DF\left(F_{\mu_\infty}\left(\frac{\xi}{-\alpha}\right)\right)\right\}.$$

(3.30)

Repeated application of T yields

$$T^n[F_{\bar{\mu}}](\xi) \simeq T^n[F_{\mu_\infty}](\xi) + (\bar{\mu} - \mu_\infty)L\{T^{n-1}[F_{\mu_\infty}]\}$$
$$[L\{T^{n-2}[F_{\mu_\infty}]\}[\cdots[L\{F_{\mu_\infty}\}[DF]]\cdots]](\xi). \qquad (3.31)$$

Using (3.25) in the limit as $n \to \infty$, (3.31) becomes

$$T^n[F_{\bar{\mu}}](\xi) \simeq g(\xi) + (\bar{\mu} - \mu_\infty)L^n\{g\}[DF](\xi). \qquad (3.32)$$

Denote the set of eigenfunctions of $L\{g\}$ as $\{\psi_i\}$ with $L\{g\}[\psi_i] = \lambda_i\psi_i$, where only one of the λ_i's, say λ_1, is assumed to be greater than 1. We further assume that the function $DF(\xi)$ can be expanded in terms of the $\{\psi_i\}$ according to

$$DF(\xi) = \sum_i c_i\psi_i(\xi). \qquad (3.33)$$

Concerning the validity of these conjectures see Feigenbaum (1979) and Collet et al. (1980). Then for large n

$$L^n\{g\}[DF](\xi) \simeq c_1\lambda_1^n\psi_1(\xi), \qquad (3.34)$$

where λ_1 is the eigenvalue greater than 1 and c_1 is constant. Inserting in (3.32) gives

$$T^n[F_{\bar{\mu}_n}](\xi) \simeq g(\xi) + (\bar{\mu}_n - \mu_\infty)c_1\lambda_1^n\psi_1(\xi), \qquad (3.35)$$

where we have set $\bar{\mu} = \bar{\mu}_n$. Since $F_{\bar{\mu}_n}(0) = 0$, we see that $T^n[F_{\bar{\mu}_n}](0) = (-\alpha)^n F_{\bar{\mu}_n}^{2^n}(0) = 0$, and thus (3.35) becomes

$$0 \simeq 1 + (\bar{\mu}_n - \mu_\infty)c_1\lambda_1^n\psi_1(0)$$

Consequently,

$$\mu_\infty - \bar{\mu}_n = \frac{\text{constant}}{\lambda_1^n} \quad \text{for } n \gg 1. \qquad (3.36)$$

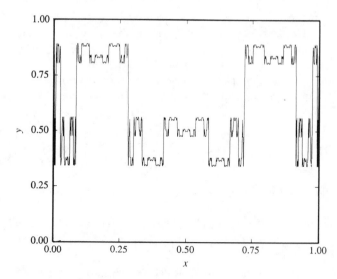

FIGURE 3.10   The 2^6 iterate of $F_{\bar{\mu}_6}$ for $\mu = 3.57$.

Comparing with (2.21), we conclude that $\lambda_1 \equiv \delta$. This follows since all sequences of μ determined by the period-doubling sequence must approach μ_∞ with the same geometric convergence rate. In particular the supercycle values $\bar{\mu}_1, \bar{\mu}_2, \ldots$ converge to μ_∞. We see that δ is just that unique eigenvalue of $L\{g\}$ greater than 1 and measures how fast "points" near to g in function space diverge away from g under action of T.

We can use this result to calculate an estimate of δ.

$$L\{g\}[\psi_1](0) = \delta\psi_1(0) \tag{3.37}$$

and from (3.30) and after using (3.12) this becomes

$$-\alpha[g'(1)\psi_1(0) + \psi_1(1)] = \delta\psi_1(0). \tag{3.38}$$

We need $g'(1)$, and this can be obtained by differentiating twice the fixed point equation (3.11) for g and by using the facts $g'(0) = 0$ and $g''(0) \neq 0$, appropriate for a quadratic maximum. We obtain $g'(1) = -\alpha$, and substitute to find

$$-\alpha[-\alpha\psi_1(0) + \psi_1(1)] = \delta\psi_1(0). \tag{3.39}$$

Assuming the lowest order approximation for ψ_1, namely, that it is a constant, gives from (3.39)

$$\alpha^2 - \alpha = \delta.$$

Using $\alpha = 2.503$, we obtain $\delta = 3.762$, which is off by 19% from the Feigenbaum number 4.6692.

3.2 Scaling and Self-Similarity in the Logistic Map

Figure 3.10 shows a high-order iterate of a logistic map such that when suitably magnified by six powers of α, it gives a high-order approximation to the universal function $g_0(\xi)$ near $\xi = 0$ or $x = \frac{1}{2}$. One cannot avoid being struck by the appearance of battlements within battlements. This recurrence of the same patterns on ever finer scales is referred to as *self-similar* structure. The limit of the function iterates $F_{\bar{\mu}_n}^{2^n}(x)$ would truely give a function with self-similar structure. No matter how fine the scale, it would always look the same—battlements within battlements. This self-similar form also shows that we need not focus on $\xi = 0$ to perform the scale expansion and magnification in order to get the universal functions. Any self-similar point would do.

A second example of self-similarity is evident in Fig. 3.8 where the position of the elements in the 2^n-supercycles are plotted for increasing values of n. For large n, as one magnifies any region of the interval $(0,1)$ containing cycle elements, the structure in the magnified region is the same as the structure before magnification. We examine the self-similarity in the position of these cycle elements in greater detail.

In (3.5) and (3.6) we showed the relations

$$d_n(0) = F_{\bar{\mu}_n}^{2^{n-1}}(0) \tag{3.40}$$

and

$$\frac{d_{n+1}(0)}{d_n(0)} \simeq -\frac{1}{\alpha}, \qquad n \gg 1, \tag{3.41}$$

where we have added an argument to the distance d_n to denote that we are referring to the cycle element at $\xi = 0$. We generalize this concept to the ratio of the distance between all nearest neighbors in a 2^n-supercycle. We denote the elements in such a 2^n-supercycle as

$$0 = \xi_0 \rightarrow \xi_1 \rightarrow \cdots \rightarrow \xi_i \rightarrow \xi_{i+1} \rightarrow \cdots \xi_{2^n-1} \rightarrow \xi_{2^n} = \xi_0 = 0. \tag{3.42}$$

We recall that the way to find the element in a 2^n-supercycle nearest to say ξ_i is by iterating 2^{n1} times on this point with the function $F_{\bar{\mu}_n}$, where $\mu_n < \bar{\mu}_n < \mu_{n+1}$. Thus

$$d_n(i) = \xi_i - F_{\bar{\mu}_n}^{2^{n-1}}(\xi_i). \tag{3.43}$$

Note that the sign change from (3.40) is unimportant since the distances in (3.40) alternate in sign with n. The quantity $d_n(i)$ in (3.43) gives the distance between the cycle element ξ_i and its nearest neighbor in a 2^n-supercycle. We desire to know how these distances scale for all i. Thus we focus on

$$\sigma_n(i) \equiv \frac{d_{n+1}(i)}{d_n(i)}. \tag{3.44}$$

Note that the cycle elements for computing $d_{n+1}(i)$ and $d_n(i)$ are different and that there are twice as many elements in the 2^{n+1}-supercycle as in the 2^n-supercycle. Furthermore, since the cycle elements get closer with increasing n, it must be the case that $|\sigma_n(i)| < 1$. We already know for $i = 0$ that

$$\sigma_n(0) \simeq -\frac{1}{\alpha}. \tag{3.45}$$

Furthermore, since the point nearest $\xi_0 = 0$, namely $\xi_{2^{n-1}}$, is close to the quadratic maximum, we can show that under one iteration of the map the ratio $\sigma_n(i)$ must go to $(1/\alpha)^2$. Consider $\xi_0 = 0$ and its nearest neighbor $F_{\bar{\mu}_n}^{2^{n-1}}(\xi_0) = \xi_0 \pm \epsilon$. From a Taylor's expansion

$$F_{\bar{\mu}_n}(\xi_0 \pm \epsilon) - F_{\bar{\mu}_n}(\xi_0) \simeq \pm \frac{1}{2} F_{\bar{\mu}_n}''(\xi_0)\epsilon^2. \tag{3.46}$$

Since ϵ scales like $(-1/\alpha)$, we see that

$$\sigma_n(1) \simeq \frac{1}{\alpha^2}. \tag{3.47}$$

We now obtain two useful relations between distances in the cycle.

$$d_{n+1}(i + 2^n) = F_{\bar{\mu}_{n+1}}^{2^n}(\xi_i) - F_{\bar{\mu}_{n+1}}^{2^n}\left(F_{\bar{\mu}_{n+1}}^{2^n}(\xi_i)\right) = F_{\bar{\mu}_{n+1}}^{2^n}(\xi_i) - F_{\bar{\mu}_{n+1}}^{2^{n+1}}(\xi_i)$$
$$= F_{\bar{\mu}_{n+1}}^{2^n}(\xi_i) - \xi_i = -d_{n+1}(i), \tag{3.48}$$

since ξ_i must be an element of a 2^{n+1}-supercycle for computing d_{n+1} and is therefore a fixed point of $F_{\bar{\mu}_{n+1}}^{2^{n+1}}$. Similarly, for the other supercycle

$$d_n(i + 2^n) = F_{\bar{\mu}_n}^{2^n}(\xi_i) - F_{\bar{\mu}_n}^{2^{n-1}}\left(F_{\bar{\mu}_n}^{2^n}(\xi_i)\right) = \xi_i - F_{\bar{\mu}_n}^{2^{n-1}}(\xi_i) = d_n(i), \tag{3.49}$$

which is obvious when we recall that $\xi_i = \xi_{i+2^n}$ for a 2^n-supercycle. As a consequence

$$\sigma_n(i + 2^n) = \frac{d_{n+1}(i + 2^n)}{d_n(i + 2^n)} = \frac{-d_{n+1}(i)}{d_n(i)} = -\sigma_n(i). \tag{3.50}$$

In particular for $i = 0$

$$\sigma_n(2^n) = -\sigma_n(0) = \frac{1}{\alpha}. \tag{3.51}$$

We now go back to the definition (3.43) for $d_n(i)$ and focus on some special values of i. We call them i_m, where $i_m = 2^{n-m} = 2^n, \ldots, 8, 4, 2, 1$ for $m = 0, 1, \ldots, n$. This labeling picks out $n + 1$ elements of a 2^n element supercycle.

$$d_n(i_m) = d_n(2^{n-m}) = F_{\bar{\mu}_n}^{2^{n-m}}(0) - F_{\bar{\mu}_n}^{2^{n-1}}\left(F_{\bar{\mu}_n}^{2^{n-m}}(0)\right). \tag{3.52}$$

From (3.16) we can write

$$F_{\bar{\mu}_n}^{2^{n-m}}(0) \simeq (-\alpha)^{m-n} g_m(0), \qquad n \gg 1. \tag{3.53}$$

Using this result and (3.16) directly, we have

$$
\begin{aligned}
F_{\bar{\mu}_n}^{2^{n-1}}\left(F_{\bar{\mu}_n}^{2^{n-m}}(0)\right) = F_{\bar{\mu}_n}^{2^{n-m}}\left(F_{\bar{\mu}_n}^{2^{n-1}}(0)\right) &\simeq F_{\bar{\mu}_n}^{2^{n-m}}\left((-\alpha)^{1-n} g_1(0)\right) \\
&= F_{\bar{\mu}_n}^{2^{n-m}}\left((-\alpha)^{1-n}(-\alpha)^{n-m} \frac{g_1(0)}{(-\alpha)^{n-m}}\right) \\
&\simeq (-\alpha)^{m-n} g_m\left((-\alpha)^{1-m} g_1(0)\right).
\end{aligned}
\tag{3.54}
$$

Consequently, the distance can be expressed in the form

$$d_n(i_m) \simeq (-\alpha)^{m-n} g_m(0) - (-\alpha)^{m-n} g_m\left((-\alpha)^{1-m} g_1(0)\right).$$

In a similar fashion we obtain

$$d_{n+1}(i_m) \simeq (-\alpha)^{m-n} g_{m+1}(0) - (-\alpha)^{m-n} g_{m+1}\left((-\alpha)^{-m} g_1(0)\right).$$

We take the ratio of these distances to find

$$\sigma_n(i_m) \simeq \frac{g_{m+1}(0) - g_{m+1}\left((-\alpha)^{-m} g_1(0)\right)}{g_m(0) - g_m\left((-\alpha)^{1-m} g_1(0)\right)}.$$

This result is exact in the limit $n \to \infty$, and we define

$$\sigma(i_m) = \lim_{n \to \infty} \sigma_n(i_m).$$

It is convenient to use a new variable $t_m = i_m/2^{n+1} = 1/2^{m+1}$. Recall that for $i = 2^n$, $\xi_i = \xi_{2^n}$, and this equals 0 for a 2^n-supercycle. Thus $m = 0$ corresponds to ξ_{2^n} or $t_0 = \frac{1}{2}$. Similarly, for $m = n$, $t_m = 1/2^{n+1}$ and in the limit that $n \to \infty$, $t_n \to 0$. Thus all the special points in the supercycle are labeled as the reciprocal of some power of 2 on the real line between 0 and $\frac{1}{2}$. They are particular elements in the binary representation in (2.23). Thus

$$\sigma(t_m) = \frac{g_{m+1}(0) - g_{m+1}\left((-\alpha)^{-m} g_1(0)\right)}{g_m(0) - g_m\left((-\alpha)^{1-m} g_1(0)\right)}. \tag{3.55}$$

Since the g's are all universal, it follows that $\sigma(t_m)$ is universal.

Now we compute some values for $\sigma(t)$. First we look at $t_0 = \frac{1}{2}$.

$$\sigma\left(\frac{1}{2}\right) = \frac{g_1(0) - g_1\left(g_1(0)\right)}{g_0(0) - g_0\left(-\alpha g_1(0)\right)}.$$

From (3.16)

$$g_0(\xi) = \lim_{n \to \infty} (-\alpha)^n F_{\bar{\mu}_n}^{2^n}(\xi/(-\alpha)^n).$$

Since $\xi_0 = 0$ is the first element in the 2^n-supercycle, $F_{\bar{\mu}_n}^{2^n}(0) = 0$ and thus $g_0(0) = 0$. Furthermore, from (3.21) we have $g_1(g_1(0)) = 0$. Also from (3.21)

$$g_0\big(-\alpha g_1(0)\big) = -\alpha g_1\Big(g_1\big(-\alpha g_1(0)/(-\alpha)\big)\Big) = -\alpha g_1\Big(g_1\big(g_1(0)\big)\Big) = -\alpha g_1(0).$$

Substituting these results gives

$$\sigma\left(\frac{1}{2}\right) = \frac{1}{\alpha}. \tag{3.56}$$

This result is clearly consistent with (3.51).

Another easy value to obtain is $\sigma(t_m)$ for $m \to \infty$. In this limit all universal functions go to g of (3.22) and (3.23).

$$\sigma(0^+) = \lim_{m \to \infty} \frac{g(0) - g\big((-\alpha)^{-m}g_1(0)\big)}{g(0) - g\big((-\alpha)^{1-m}g_1(0)\big)}. \tag{3.57}$$

We know that $g(0) = 1$, and for small values of the argument

$$g(\xi) \simeq 1 + \frac{1}{2}g''(0)\xi^2.$$

Substitution into (3.57) gives

$$\sigma(0^+) = \frac{1}{\alpha^2}. \tag{3.58}$$

For $m = 1$, $t_1 = \frac{1}{4}$ and

$$\sigma\left(\frac{1}{4}\right) = \big[g_2(0) - g_2\big(g_1(0)/(-\alpha)\big)\big]/g_1(0). \tag{3.59}$$

In order to evaluate $\sigma(\frac{1}{4})$ we need a numerical computation of the universal functions in (3.59). Feigenbaum (1980b) gives

$$\sigma(\frac{1}{4}) = 0.1752\ldots. \tag{3.60}$$

Since $m = 1$ corresponds to $\xi^{2^{n-1}}$ in the sequence of (3.42), the value in (3.60) represents a sudden change in scaling half way around the cycle.

The result contained in (3.55) gives us the scaling of the distances between nearest neighbors for some of the elements in the cycle represented by $t_m = 1/2^{m+1}$. But there are still other elements in the supercycle besides those special ones in the list following (3.51). Since σ is a discontinuous function and to get values just larger than those obtained, we consider

$$d_n(i_m + 1) = F_{\bar{\mu}_n}\left(F_{\bar{\mu}_n}^{2^{n-m}}(0)\right) - F_{\bar{\mu}_n}^{2^{n-1}}\left(F_{\bar{\mu}_n}\left(F_{\bar{\mu}_n}^{2^{n-m}}(0)\right)\right).$$

Commuting these maps and using (3.53) and (3.54) gives

$$d_n(i_m + 1) \simeq F_{\bar{\mu}_n}\left((-\alpha)^{m-n} g_m(0)\right)$$
$$- F_{\bar{\mu}_n}\left((-\alpha)^{m-n} g_m\left((-\alpha)^{1-m} g_1(0)\right)\right). \qquad (3.61)$$

As $n \to \infty$, the arguments of $F_{\bar{\mu}_n}$ in (3.61) become small, and so it is appropriate to expand about 0. Noting that $F'_{\bar{\mu}_n}(0) = 0$, we find

$$d_n(i_m + 1) \simeq \frac{1}{2} F''_{\bar{\mu}_n}(0)(\alpha^2)^{m-n}\left\{ [g_m(0)]^2 - [g_m\left((-\alpha)^{1-m} g_1(0)\right)]^2 \right\}.$$

In similar fashion

$$d_{n+1}(i_m + 1) \simeq \frac{1}{2} F''_{\bar{\mu}_{n+1}}(0)(\alpha^2)^{m-n}\left\{ [g_{m+1}(0)]^2 - [g_{m+1}\left((-\alpha)^{-m} g_1(0)\right)]^2 \right\}.$$

As $n \to \infty$, $\bar{\mu}_n$ and $\bar{\mu}_{n+1} \to \mu_\infty$ and thus

$$\sigma(i_m + 1) = \frac{[g_{m+1}(0)]^2 - [g_{m+1}\left((-\alpha)^{-m} g_1(0)\right)]^2}{[g_m(0)]^2 - [g_m\left((-\alpha)^{1-m} g_1(0)\right)]^2}. \qquad (3.62)$$

Dividing $(i_m + 1)$ by 2^{n+1} to find the corresponding t label, we get $t = 1/2^{m+1} + 1/2^{n+1}$. As $n \to \infty$ this goes to

$$\sigma\left(\frac{1}{2^{m+1}} + 0^+\right) = \frac{[g_{m+1}(0)]^2 - [g_{m+1}\left((-\alpha)^{-m} g_1(0)\right)]^2}{[g_m(0)]^2 - [g_m\left((-\alpha)^{1-m} g_1(0)\right)]^2}. \qquad (3.63)$$

Special cases are for $m = 0$, $m = 1$, and $m \to \infty$ and are computed as was done for $\sigma(1/2^{m+1})$. We find

$$\sigma\left(\frac{1}{2} + 0^+\right) = -\frac{1}{\alpha^2}, \qquad \sigma(0^+) = \frac{1}{\alpha^2}.$$

Feigenbaum (1980b) gives $\sigma(\frac{1}{4} + 0^+) = 0.4191\ldots$. The values for $\sigma(\frac{1}{2} + 0^+)$ and $\sigma(0^+)$ can also be obtained from the limit as $n \to \infty$ of (3.50), which becomes for the special cycle elements considered

$$\sigma\left(\frac{1}{2^{m+1}} + \frac{1}{2}\right) = -\sigma\left(\frac{1}{2^{m+1}}\right). \qquad (3.64)$$

For $m = 0$ this implies $\sigma(1) = -\sigma(\frac{1}{2}) = -\frac{1}{\alpha}$.

Having computed the largest discontinuites we can now get σ on any rational value of t, and hence any supercycle element. We simply write its binary representation and then look at $i = 2^{n-m_1} + 2^{n-m_2} + \ldots$. For example, $i = 3 = 2^{n-m_1} + 2^{n-m_2}$ with $m_1 = n$ and $m_2 = n - 1$. Then make use of the fact that functional composition commutes. By the same methods employed previously, we find

$$d_n(2^{n-m_1} + 2^{n-m_2}) \simeq (-\alpha)^{m_2-n} g_{m_2}((-\alpha)^{m_1-m_2} g_{m_1}(0))$$

$$- (-\alpha)^{m_2-n} g_{m_2}\left((-\alpha)^{m_1-m_2} g_{m_1}\left(\frac{g_1(0)}{(-\alpha)^{1-m_1}}\right)\right).$$

Similarly for $d_{n+1}(2^{n-m_1} + 2^{n-m_2})$ implying

$$\sigma\left(\frac{1}{2^{m_1+1}} + \frac{1}{2^{m_2+1}}\right)$$

$$= \left[g_{m_2+1}\left(\frac{g_{m_1+1}(0)}{(-\alpha)^{m_2-m_1}}\right) - g_{m_2+1}\left((-\alpha)^{m_1-m_2} g_{m_1+1}\left(\frac{g_1(0)}{(-\alpha)^{-m_1}}\right)\right)\right]$$

$$\left/ \left[g_{m_2}\left(\frac{g_{m_1}(0)}{(-\alpha)^{m_2-m_1}}\right) - g_{m_2}\left((-\alpha)^{m_1-m_2} g_{m_1}\left(\frac{g_1(0)}{(-\alpha)^{1-m_1}}\right)\right)\right].\right. \quad (3.65)$$

In general for $i = 2^{n-m_1} + 2^{n-m_2} + \cdots$ one finds

$$\sigma\left(\frac{1}{2^{m_1+1}} + \frac{1}{2^{m_2+1}} + \cdots\right) = \left[\left[\cdots g_{m_2+1}\left(\frac{g_{m_1+1}}{(-\alpha)^{m_2-m_1}}\right)\cdots\right)\right]$$

$$- \left[\cdots g_{m_2+1}\left((-\alpha)^{m_1-m_2} g_{m_1+1}\left(\frac{g_1(0)}{(-\alpha)^{-m_1}}\right)\right)\cdots\right)\right]\right]$$

$$\left/ \left[\left[\cdots g_{m_2}\left(\frac{g_{m_1}(0)}{(-\alpha)^{m_2-m_1}}\right)\cdots\right)\right]\right.$$

$$- \left[\cdots g_{m_2}\left((-\alpha)^{m_1-m_2} g_{m_1}\left(\frac{g_1(0)}{(-\alpha)^{1-m_1}}\right)\right)\cdots\right)\right]\right]. \quad (3.66)$$

For the discontinuity we simply add 1 to i, giving $i = 2^{n-m_1} + 2^{n-m_2} + \cdots + 1$ and resulting in

$$\sigma\left(\frac{1}{2^{m_1+1}} + \frac{1}{2^{m_2+1}} + \cdots + 0^+\right) = \frac{[\cdots]^2 - [\cdots]^2}{[\cdots]^2 - [\cdots]^2}, \quad (3.67)$$

where the contents of the brackets are the same as in (3.66). Thus we see that the function σ is universal and discontinuous at all rationals. However, with an increasing number of terms in the binary expansion of a rational the discontinuity becomes milder. Figure 3.11 taken from Feigenbaum (1980b) plots σ^{-1} and shows clearly the symmetry about $\frac{1}{2}$ given by

$$\sigma\left(\frac{1}{2m_1 + 1} + \frac{1}{2^{m_2+1}} + \cdots + \frac{1}{2}\right) = -\sigma\left(\frac{1}{2^{m_1+1}} + \frac{1}{2^{m_2+1}} + \cdots\right) \quad (3.68)$$

of which (3.64) is an example. Furthermore, we see that this function is self-similar.

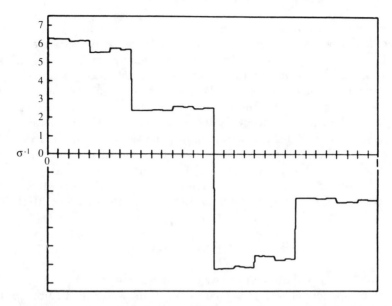

FIGURE 3.11   The scaling function for distances between cycle elements according to Feigenbaum (1980b).

This scaling function encodes in a universal way the manner in which distances between cycle elements scale as a period-doubling sequence of bifurcations proceeds toward chaos. Very often the absolute measurement of such quantities as distances is not accessible, whereas scalings may be. Attempts have also been made to measure the scaling function directly as reported by Belmonte et al. (1988). In the next section we study an application of the scaling function to obtain scaling properties of Fourier amplitudes of systems that undergo period doubling into chaos.

3.3 Subharmonic Scaling

Consider a system described by the differential system

$$\dot{\mathbf{x}} = \mathbf{f}_\mu(\mathbf{x}, t), \tag{3.69}$$

where μ is the mapping parameter as before. Specifically, consider period-doubling systems in which $\mathbf{x}(t)$ is periodic with period $T_n = 2^n T_0$ for $\mu_n < \mu < \mu_{n+1}$ and $\mu_n \to \mu_\infty$ as $n \to \infty$. The parameters μ_n at each n are chosen so as to have identical stability properties, as determined by the eigenvalues of the Jacobian matrix $\partial \mathbf{f}/\partial \mathbf{x}$. This dependence of stability on eigenvalues is considered more directly in Chapter 5. For example, the μ_n may be chosen to be those values in each parameter range corresponding to a particular attractor that gives the most rapid convergence to the attractor state.

The frequency spectrum of $\mathbf{x}(t)$ has a fundamental $\omega_n = \omega_0/2^n$ with equally spaced harmonics up to ω_0. As $n \to \infty$, $\mathbf{x}(t)$ becomes aperiodic and the spectrum becomes continuous. It is precisely this subharmonic spectrum that we examine in detail in this section. We make the assumption throughout that the fundamental frequency ω_0 does not change in this process as $n \to \infty$.

At $\mu = \mu_n$, $\mathbf{x}(t_0 + T_n) = \mathbf{x}(t_0)$, and the motion $\mathbf{x}(t)$ can be divided into 2^n parts or "cycles" that are roughly similar, each of approximately T_0 in duration. When $\mu = \mu_{n+1}$, $\mathbf{x}(t_0 + T_n)$ no longer equals $\mathbf{x}(t_0)$, but almost does. It requires an additional 2^n cycles of duration T_0 before it does, i.e., another cycle of length T_n is required before $\mathbf{x}(t)$ comes back to $\mathbf{x}(t_0)$. Thus we focus attention on

$$d_n(t) = x^{(n)}(t) - x^{(n)}(t + T_{n-1}), \tag{3.70}$$

where for brevity we have dropped a vector index and regard (3.70) as holding for each component of $\mathbf{x}(t)$. The quantity $d_n(t)$ measures the failure of $\mathbf{x}(t)$ to equal itself after one-half of its true period. Note that $d_n(t + T_n) = d_n(t)$ and $d_n(t + T_{n-1}) = -d_n(t)$. Now assume that $d_{n+1}(t)$ is proportional to $d_n(t)$, i.e., the way $d_{n+1}(t)$ fails to vanish is scaled to the way that $d_n(t)$ fails to vanish.

$$d_{n+1}(t) \simeq \sigma(t/T_{n+1})d_n(t). \tag{3.71}$$

The argument of σ ensures that it has the same periodicity that $d_{n+1}(t)$ has. Because $d_n(t)$ is periodic with half the period of $d_{n+1}(t)$, we see that $d_{n+1}(t)$ is made up of two scaled copies of $d_n(t)$. From the foregoing periodicity conditions we obtain

$$\sigma(\tau + 1) = \sigma(\tau),$$
$$\sigma(\tau + \frac{1}{2}) = -\sigma(\tau) \tag{3.72}$$

and since

$$\sigma\left(\frac{2t}{T_{n+1}}\right) = \sigma\left(\frac{t}{T_n}\right),$$

the ratio of the distance functions $r_n(t)$ satisfies

$$r_{n+1}(2t) \equiv \frac{d_{n+1}(2t)}{d_n(2t)} \simeq \sigma\left(\frac{2t}{T_{n+1}}\right) = \sigma\left(\frac{t}{T_n}\right) \simeq \frac{d_n(t)}{d_{n-1}(t)} = r_n(t). \tag{3.73}$$

So the ratio of the distance functions $r_n(\tau)$ plotted against a scaled time τ with $\tau = 1$ corresponding to $t = T_n$ should be identical and equal the function

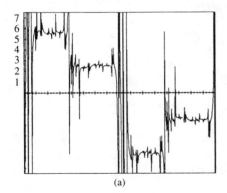
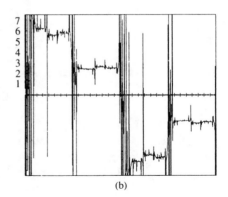

(a)   (b)

FIGURE 3.12 The σ function obtained numerically from Duffing's equation by plotting the ratios d_4/d_3 and d_5/d_4. The horizontal scale is compressed by a factor of 2 in (b) from that of (a). After Feigenbaum (1980a).

$\sigma(\tau)$. Figure 3.12 from Feigenbaum (1980a) shows these ratios for $n = 3, 4$ and $n = 4, 5$ in the case that (3.69) is given by Duffing's equation:

$$\dot{x} = y,$$
$$\dot{y} = -\mu y - x^3 + b \sin 2\pi t.$$

The μ_n's are determined by choosing the value of μ giving the most rapid rate of convergence to the 2^n-cycle in the basin of attraction for this cycle.

There is a large class of systems with dynamics determined by a differential equation of the type in (3.69) that undergo the period doubling sequence and for which similar results are obtained. We note the similarity of the curves in Fig. 3.12 with Fig. 3.11.

Having made this connection with cycle elements and trajectories in phase space, let us examine the subharmonic spectrum. We first note that the even and odd frequency components of $x^{(n)}(t)$ are fundamentally different for the following reason: The odd harmonics at μ_{n+1} are all absent in the spectrum at μ_n. This follows because at μ_n the fundamental is $\omega_n = \omega_0/2^n$ and at μ_{n+1} the fundamental is $\omega_{n+1} = \omega_0/2^{n+1}$. Thus $\omega_n = 2\omega_{n+1}$. At μ_{n+1} a set of new components to the spectrum of frequencies is added, namely $\omega_0/2^{n+1}, 3\omega_0/2^{n+1}, \ldots$, where all the even ones are the same as were present at μ_n. Thus to a first approximation the even components of $x^{(n+1)}(t)$ are just all the old components of $x^{(n)}(t)$ at μ. The main task in determining the spectrum at μ_{n+1} is the computation of the odd components.

By definition the kth Fourier component of $x^{(n)}(t)$ is given by

$$x_k^{(n)} = \frac{1}{T_n} \int_0^{T_n} x^{(n)}(t) e^{-2\pi i k t/T_n}\, dt, \tag{3.74}$$

where

$$x^{(n)}(t) = \sum_k x_k^{(n)} e^{2\pi ikt/T_n}. \tag{3.75}$$

We split the range of integration in two pieces:

$$x_k^{(n)} = \frac{1}{T_n} \int_0^{T_{n-1}} \cdots + \frac{1}{T_n} \int_{T_{n-1}}^{T_n} \cdots$$

and then shift the variable of integration in the second integral to obtain

$$x_k^{(n)} = \frac{1}{2T_{n-1}} \int_0^{T_{n-1}} dt \, [x^{(n)}(t) + (-1)^k x^{(n)}(t + T_{n-1})] e^{-\pi ikt/T_{n-1}}. \tag{3.76}$$

Consider the even harmonics of $x^{(n+1)}(t)$ first. From (3.76)

$$
\begin{aligned}
x_{2k}^{(n+1)} &= \frac{1}{2T_n} \int_0^{T_n} dt \, [x^{(n+1)}(t) + x^{(n+1)}(t + T_n)] e^{-2\pi ikt/T_n} \\
&= \frac{1}{2T_n} \int_0^{T_n} dt \, [2x^{(n+1)}(t) - d_{n+1}(t)] e^{-2\pi ikt/T_n},
\end{aligned}
$$

and $d_{n+1}(t)$ will be small, particularly so for large n. Thus to a first approximation $d_{n+1}(t) \approx 0$ and $x^{(n+1)}(t) \approx x^{(n)}(t)$ giving

$$x_{2k}^{(n+1)} \simeq \frac{1}{T_n} \int_0^{T_n} dt \, x^{(n)}(t) e^{-2\pi ikt/T_n} = x_k^{(n)}.$$

Recall that the fundamental for $x^{(n+1)}(t)$ is $\omega_0/2^{n+1}$ so that the $2k$th harmonic has frequency $k\omega_0/2^n$, which is the kth harmonic of $x^{(n)}(t)$. The power in the kth frequency is proportional to $|x_k^{(n)}|^2$ and we see that $|x_{2k}^{(n+1)}|^2 \simeq |x_k^{(n)}|^2$.

Now we turn to the odd harmonics. From (3.76)

$$
\begin{aligned}
x_{2k+1}^{(n+1)} &= \frac{1}{2T_n} \int_0^{T_n} dt \, [x^{(n+1)}(t) - x^{(n+1)}(t + T_n)] e^{-\pi i(2k+1)t/T_n} \\
&= \frac{1}{2T_n} \int_0^{T_n} dt \, d_{n+1}(t) e^{-\pi i(2k+1)t/T_n} \\
&\simeq \frac{1}{2T_n} \int_0^{T_n} dt \, \sigma(t/2T_n) d_n(t) e^{-\pi i(2k+1)t/T_n}.
\end{aligned}
\tag{3.77}
$$

From (3.70) and (3.75) find that

$$
\begin{aligned}
d_n(t) &= \sum_k x_k^{(n)} [e^{2\pi ikt/T_n} - e^{2\pi ik(t+T_{n-1})/T_n}] \\
&= \sum_k x_k^{(n)} e^{2\pi ikt/T_n} (1 - e^{\pi ik}) \\
&= \sum_k x_k^{(n)} e^{2\pi ikt/T_n} (1 - (-1)^k).
\end{aligned}
$$

So only terms with k odd contribute to the sum resulting in

$$d_n(t) = 2 \sum_m x_{2m+1}^{(n)} e^{2\pi i (2m+1)t/T_n}.$$

Substituting into (3.77) gives

$$x_{2k+1}^{(n+1)} \simeq \sum_m x_{2m+1}^{(n)} \frac{1}{T_n} \int_0^{T_n} dt\, \sigma(t/2T_n) e^{2\pi i t(2m+1-(2k+1)/2)/T_n}.$$

Now change the variable to $\zeta = t/T_n$ to find

$$x_{2k+1}^{(n+1)} \simeq \sum_m x_{2m+1}^{(n)} \int_0^1 d\zeta\, \sigma(\zeta/2) e^{2\pi i(2m+1-(2k+1)/2)\zeta}. \qquad (3.78)$$

Thus a knowlege of the function σ, i.e., the scaling law, and the spectrum of $x^{(n)}(t)$ allows the computation of the spectrum of $x^{(n+r)}(t)$ for all $r \geq 1$.

From the results of the previous section and as seen in Fig. 3.11, σ can be approximated by the function

$$\sigma(\zeta/2) \simeq \begin{cases} \frac{1}{\alpha^2}, & 0 < \zeta < \frac{1}{2} \\ \frac{1}{\alpha}, & \frac{1}{2} < \zeta < 1 \end{cases}.$$

With this approximation the integral in (3.78) becomes

$$\frac{1}{\alpha^2} \int_0^{1/2} d\zeta\, e^{2\pi i (2m+1-(2k+1)/2)\zeta} + \frac{1}{\alpha} \int_{1/2}^1 d\zeta\, e^{2\pi i(2m+1-(2k+1)/2)\zeta}$$

$$= -\left[\left(\frac{1}{\alpha} + \frac{1}{\alpha^2} \right) + i(-1)^k \left(\frac{1}{\alpha} - \frac{1}{\alpha^2} \right) \right] / 2\pi i (2m+1-(2k+1)/2).$$

Thus

$$x_{2k+1}^{(n+1)} = \frac{i}{2\pi} \left[\left(\frac{1}{\alpha} + \frac{1}{\alpha^2} \right) + i(-1)^k \left(\frac{1}{\alpha} - \frac{1}{\alpha^2} \right) \right] \sum_m x_{2m+1}^{(n)} / (2m+1-(2k+1)/2).$$

We make an integral approximation to the preceding sum over m.

$$\sum_m x_{2m+1}^{(n)} / (2m+1-(2k+1)/2)$$

$$\rightarrow \frac{1}{2} P \int \frac{x^{(n)}(m)}{m - k/2}\, dm \qquad (3.79)$$

$$\simeq -\frac{\pi i}{2} x_{k/2}^{(n)},$$

where the last result follows from extending a contour integral into the lower half-plane. That no contribution is made by this extension follows from (3.74) which shows that the contribution to the Fourier amplitudes from a contour at ∞ in the lower half-plane is zero. Consequently, for the odd amplitudes

$$|x_k^{(n+1)}| \simeq \frac{1}{4}\left[\left(\frac{1}{\alpha}+\frac{1}{\alpha^2}\right)^2 + \left(\frac{1}{\alpha}-\frac{1}{\alpha^2}\right)^2\right]^{\frac{1}{2}}|x_{k/2}^{(n)}| \equiv \gamma|x_{k/2}^{(n)}|. \qquad (3.80)$$

The meaning of this result is that one can estimate the Fourier amplitudes of the spectrum at the $(n+1)$ cycle from the spectrum at the n cycle. To do so, smoothly interpolate between neighboring elements in the n cycle to obtain $|x_{k/2}^{(n)}|$, then reduce by the factor $\gamma = 0.1525$ to obtain $|x_k^{(n+1)}|$.

Thus we see at each level in the period-doubling cascade that the amplitudes of the even harmonics remain approximately the same from one level to the next, but that the amplitude of the new odd harmonics appearing after the bifurcation are down in amplitude by a factor of $\simeq 0.153$.

3.4 Experimental Comparisons

Very detailed results have been obtained in the previous section and before leaving the topic of universality in period doubling, a few comparisons with experimental observations are in order. We first summarize the characteristics of the period-doubling route to chaos.

1. As one varies a control parameter, the system goes through a sequence of stable oscillations. The period of these oscillations doubles as one goes from one stable state to the next.

2. These transitions (bifurcations) take place to a new stable state with twice the period, at parameter values that become ever closer as the ratio of their differences approaches $\delta = 4.669\ldots$ [cf. Eq. (2.21)].

3. There is a corresponding subharmonic cascade in the frequency components comprising the spectrum of the oscillation as the control parameter is varied. The fundamental frequency of the system is reduced by a factor of 2 with each bifurcation, and the number of subharmonics present doubles.

4. From one bifurcation to the next, the amplitudes of the common subharmonics remains approximately constant. The amplitudes of the new subharmonics, introduced at each bifurcation, is reduced by a factor of $\gamma = 0.1525$ or $10\log_{10}\gamma^{-1} = 8.17\ dB$ from the amplitude of the previously existing harmonics. Experimentalists have sometimes used the ratio $R = \sqrt{2}/2\gamma$.

Table 3.1 summarizes the results from numerous experiments wherein period-doubling plays a role and is adapted from Cvitanović (1984).

Table 3.1: Experimental Measurements on Period Doublings				
Experiment	Number of period doublings	δ (4.669)	α (2.503)	R (4.648)
Hydrodynamic :				
Water [1]	2			
Water [2]	4	4.3 ± 0.8		4 ± 1
Helium [3]	4	3.5 ± 0.15		$4 \pm ?$
Mercury [4]	4	4.4 ± 0.1		5 ± 1
Electronic :				
Diode [5]	4	4.5 ± 0.6		$6 \pm ?$
Diode [6]	5	4.3 ± 0.1	2.4 ± 0.1	Consistent
Transistor [7]	4	4.7 ± 0.3		Consistent
Josephson [8]	3	4.5 ± 0.3	2.7 ± 0.2	
Laser :				
Laser feedback [9]	3	4.3 ± 0.3	Consistent	
Laser [10]	2			
Laser [11]	3			
Acoustic :				
Helium [12]	3			
Helium [13]	3	4.8 ± 0.6		$6 \pm ?$
Chemical :				
B-Zh reaction [14]	3			

[1] Gollub and Benson (1980)
[2] Giglio et al. (1981)
[3] Libchaber and Maurer (1981)
[4] Libchaber et al. (1982)
[5] Linsay (1981)
[6] Testa et al. (1982)
[7] Arecchi and Lisi (1982)
[8] Yeh and Kao (1982)
[9] Hopf et al. (1981)
[10] Arecchi et al. (1982)
[11] Weiss et al. (1983)
[12] Lauterborn and Cramer (1981)
[13] Smith et al. (1982)
[14] Simoyi et al. (1982)

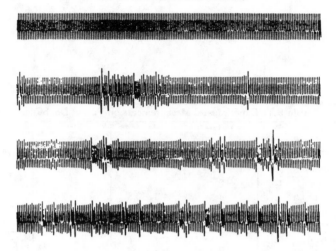

FIGURE 3.13   Transistion to intermittency in the Lorenz system. From Pomeau and Manneville (1980).

3.5 Intermittency

Even though the period-doubling route to chaos occurs frequently, it is not the only way systems can become chaotic. Another frequently occurring route is called *intermittency*. Intermittency is still under active investigation and we do not attempt a complete discussion of all known results here. For further discussion and detail we recommend Bergé et al. (1984), Schuster (1984), Pomeau and Manneville (1980), Hirsch et al. (1982), Hu and Rudnick (1982).

Suppose that one of the dynamical variables of the system is observed as a function of time. Intermittency is indicated if the system behaves in the following way. The observed variable undergoes periodic oscillations that are regular and stable. After the control parameter is changed to a nearby value, these regular periodic oscillations are interupted abruptly by "bursts." As the control parameter is further changed, the bursts occur more frequently and the duration of the regular oscillations decreases. Figure 3.13 shows a transistion to chaos through intermittency in the Lorenz system (discussed in detail in Chapter 6).

The familiar logistic map also readily displays an intermittent transistion to chaos. Figure 3.14 shows the logistic map iterates for the map parameter μ in the neighborhood of $\mu_c = 1 + 2\sqrt{2} = 3.828\ldots$, which is the value of μ for the onset of the large band evident in Fig. 2.7 for the 3-cycle. As $\epsilon = (\mu_c - \mu)$ increases from zero, the logistic map iterates make the transistion from a stable 3-cycle to chaos. For a value of μ near μ_c, where the logistic map exhibits intermittency as in the second trace of Fig. 3.14, we plot in Fig. 3.15 every third point in an iteration of the logisitic map. The stretches, on the order of 100 iterations long, where every third point comes back to the same value, show the laminar phases in this map. The chaotic bursts are the regions

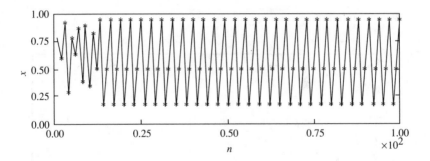

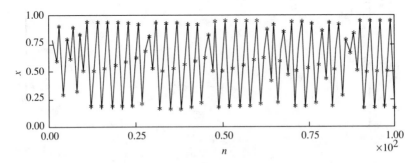

FIGURE 3.14   Transistion to intermittency in the logistic map. The upper sequence of iterates is for $\mu = 3.8304$ and the lower is for $\mu = 3.8264$.

that connect these regular phases. It is clear in Fig. 3.15 that not all laminar regions contain the same number of iterations.

The intermittency route to chaos has been further classified in three types. We focus primarily on type I intermittency but conclude this section with a discussion comparing the various types of intermittency. Type I intermittency is characterized by a Poincaré map that undergoes a fold bifurcation (cf. Section 2.4). Figure 3.16 shows a qualitative sketch of a sequence of one-dimensional maps that undergo a fold bifurcation.

Even after the bifurcation has taken place and the map looks qualitatively like the function (c) in Fig. 3.16, there is "memory" of the vanished stable fixed point. Points near the vanished fixed point stay nearby for many applications of the map. This behavior is sketched in Fig. 3.17. The movement of the point under the one-dimensional map through the channel depicted in Fig. 3.17 corresponds to the laminar phase of the oscillations in Figs. 3.13 and 3.14. In this region a number of results can be obtained. However, it is important to note that nothing can be said in general about the reinjection process. The process that brings the system point back into the laminar region depends on global features, and little can be said about it in general. However, Bergé et al. (1984) give some model maps illustrating some possibilities.

In Table 2.1 the archetype map for a fold bifurcation is given as

$$x_{n+1} = \mu + x_n + x_n^2. \tag{3.81}$$

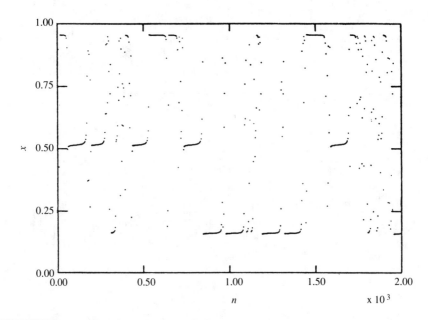

FIGURE 3.15   Laminar phases in the logistic map where every third iterate is plotted for $\mu = 3.8283$.

When μ is very small and the difference between x_n and x_{n+1} is small, we can replace the difference equation (3.81) with a differential equation:

$$\frac{dx}{dn} = \mu + x^2, \tag{3.82}$$

where we now view n as a continuous variable. It is straightforward to integrate (3.82) to obtain

$$x = \sqrt{\mu}\tan\left(\sqrt{\mu}(n - n_0)\right), \tag{3.83}$$

where the bifurcation value of x is zero, consistent with (3.81). In (3.83) we take n_0 as the value of n when x is at the narrowest part of the channel, i.e., $x = 0$. For convenience we take $n_0 = 0$. Then the right-hand side of (3.83) becomes unbounded when $\sqrt{\mu}n = \pi/2$, at which point it is clear that the differential-equation approximation to (3.81) is no longer valid. Consequently, we conclude that for the point x to move through the channel as in Fig. 3.16, the number of iterations n scales as $1/\sqrt{\mu}$. Thus, as a system undergoing intermittency approaches the bifurcation point, we expect the average duration $\langle t \rangle$ of the laminar phases to increase as $\langle t \rangle \sim 1/\sqrt{\mu}$.

Despite the specificity of this scaling result, it is difficult to measure because it requires very precise measurements near threshold ($\mu = 0$). This

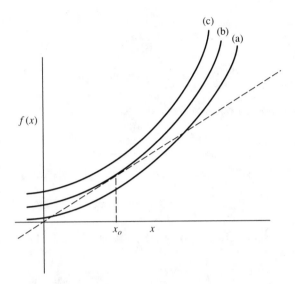

FIGURE 3.16 Sketch showing the manner in which the variation of a map parameter can lead to a fold bifurcation. $(a) \rightarrow (b) \rightarrow (c)$.

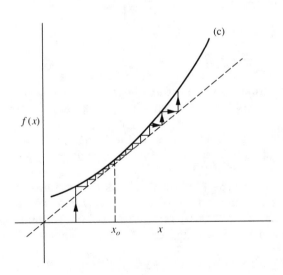

FIGURE 3.17 Qualitative sketch of the behavior of the one-dimensional map near the bifurcation point.

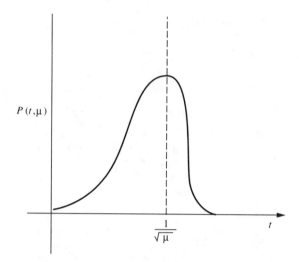

$P(t,\mu)$

$\dfrac{1}{\sqrt{\mu}}$

t

FIGURE 3.18   Qualitative sketch of the probability distribution for the duration of the laminar regions in type I intermittency.

is, of course, no problem in a numerical model and can be verified explicitly but it is difficult in experiments. Exercise 3.2 verifies this scaling for the logistic map.

The distribution in the length of the laminar phases is experimentally more accessible, however. Therefore, following Bergé et al. (1984), let us denote the probability of measuring a laminar phase with length t as $P(t,\mu)$. For μ small we know that

$$\langle t \rangle = \int_0^\infty P(t,\mu)t\,dt \simeq 1/\sqrt{\mu} \tag{3.84}$$

and that $P(t,\mu)$ has its maximum near this average value. Qualitatively, the probability function must look something like the distribution sketched in Fig. 3.18. In Exercise 3.2 the probability distribution for the logistic map is obtained.

The logistic map (2.14) serves as the archetype map leading to period doubling and through functional iteration leads us to the universal map associated with the period-doubling transistion to chaos. Similarly, (3.81) serves as the archetype for type I intermittency and through renormalization arguments leads us to a universal function for the laminar region. In Fig. 3.19, following a construction given in Schuster (1984), we obtain the scaling necessary to construct the doubling transformation analogous to (3.20) for the logistic map.

We see from the construction in Fig. 3.19 that the proper definition of the doubling operator is

$$T[f(x)] = \alpha f^2(x/\alpha). \tag{3.85}$$

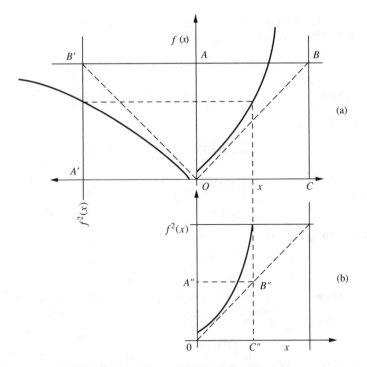

FIGURE 3.19   Qualitative sketch illustrating the construction of $f^2(x)$, where f is given in (3.81). The unit square $OABC$ is rotated by $90°$ to obtain the square $OA'B'A$. The value $f^2(x)$ is obtained by proceeding vertically from x to an intersection with the curve $f(x)$, then moving horizontally to the left to an intersection with the curve $f(x)$ in the box $OA'B'A$. Dropping vertically to the axis OA' then gives the value $f^2(x)$. Sketch (b) shows the result of this construction in the unit square. In the box $OA''B''C''$ the function $f^2(x)$ is similar to $f(x)$ in the box $OABC$.

For a fixed point of this doubling operator T in function space

$$T[f_*(x)] = \alpha f_*^2(x/\alpha) = f_*(x). \tag{3.86}$$

Equation (3.86) is very similar to (3.20), but we have different boundary conditions on the universal function f_*. Instead of (3.12) and vanishing derivative at $x = 0$, we demand

$$f_*(0) = 0, \qquad f_*'(0) = 1. \tag{3.87}$$

The solution to (3.86) and (3.87) can be obtained exactly using an auxiliary function (Hu and Rudnick, 1982). We denote this auxiliary function as $g(x)$ and define it according to

$$g\big(f(x)\big) = g(x) - a. \tag{3.88}$$

This is to say that g on the transformed point under f is related to g on the original point by a translation. Presently, no motivation for this definition of

the auxiliary function seems to be known, other than that it leads to a trick
that works. The iterated form of (3.88) gives

$$g\big(f^2(x)\big) = g\big(f(x)\big) - a = g(x) - 2a. \tag{3.89}$$

For the auxiliary function g_* corresponding to the universal function f_*,
we obtain by using (3.86) in (3.89)

$$g_*\left(\frac{1}{\alpha}f_*(\alpha x)\right) = g_*(x) - 2a. \tag{3.90}$$

By using (3.88) it is straightforward to check that (3.90) is satisfied by g_*
satisfying

$$g_*(x) = 2g_*(\alpha x). \tag{3.91}$$

As a function satisfying (3.91) we try

$$g_*(x) = x^\nu, \tag{3.92}$$

which we know will work as long as α is chosen properly. We find readily that
$\alpha = (\frac{1}{2})^{1/\nu}$. To use (3.88) to find $f_*(x)$, we need g_*^{-1}, which is clearly $x^{1/\nu}$.
Thus

$$f_*(x) = (x^\nu - a)^{1/\nu}. \tag{3.93}$$

We demand that near $x = 0$ (and of course $\mu = 0$) the expansion of (3.93)
must be (3.81) for its first two terms. Hence, $\nu = -1$, and we have

$$f_*(x) = \frac{x}{1 - x}. \tag{3.94}$$

Exercise 3.3 explains the computation of the universal function for forms near
zero more general than the archetype function (3.81).

Just as was done following (3.26) for the doubling operator associated
with the universal quadratic function, the doubling operator of (3.85) can
be linearized about the fixed point f_* or its generalization obtained in Exercise 3.3. The eigenfunction associated with the linearized operator is obtained
in Exercise 3.3 also and given as $h_\lambda(x)$. If we demand that this linearization
$h_\lambda(x)$ give a constant perturbation in function space of order $\mu \ll 1$ away
from the universal function, then in $h_\lambda(x)$ the quantity $p = 2z - 1$ and the
eigenvalue associated with this perturbation is given by $\lambda = 2^{z/(z-1)}$. We can
use this result along with a scaling argument to find the dependence of the
duration of the laminar phase on μ.

Let us denote the average number of iterations for a map to move through
the laminar region as $\langle N \rangle$. Since the doubled function $T\big[f(x)\big]$ takes only half
as many iterations for the iterated point to leave the laminar region, we have
$\langle N \rangle_0/2 = \langle N \rangle_1$ and similarly for the function doubled n times:

$$\langle N \rangle_0 = 2^n \langle N \rangle_n, \tag{3.95}$$

where the subscript on the bracket denotes the number of doublings of the map.

We also know that

$$T^n\left[f(x)\right] \simeq f_*(x) + \mu\lambda^n h(x),$$

where we assume

$$\mu\lambda^n \simeq 1 \tag{3.96}$$

for the iterated point to leave the laminar region. Taking the logarithm of (3.96) we find

$$n \simeq -\frac{\log\mu}{\log\lambda}. \tag{3.97}$$

Using (3.97) in (3.95),

$$\log\left(\frac{\langle N\rangle_0}{\langle N\rangle_n}\right) = -\left(\frac{\log\mu}{\log\lambda}\right)\log 2. \tag{3.98}$$

From the expression for the eigenvalue λ we find

$$\frac{\langle N\rangle_0}{\langle N\rangle_n} \simeq \mu^{-(z-1)/z}.$$

From (3.96) the μ for $T^n\left[f(x)\right]$ is of order 1, and thus

$$\langle N\rangle_0 \propto \mu^{-(z-1)/z}. \tag{3.99}$$

This result is clearly consistent with our former result when $z = 2$. Other values for z give different dependencies for the average length of the laminar signal, but as mentioned earlier, this dependence can be hard to determine experimentally.

Type I intermittency is associated with a fold bifurcation and the associated map becomes unstable as the derivative of the map passes through $+1$, as is summarized in Table 2.2. For high dimensional maps the derivative of the map becomes a matrix and the eigenvalues of this matrix are the relevant quantities determining stability. If the dominant eigenvalue (the one determining stability) becomes unstable by passing through $+1$, then we have type I intermittency as discussed previously. If this eigenvalue (along with its complex conjugate) crosses the unit circle in the complex plane, as the map parameter μ crosses the transition threshold, then the intermittency transition is said to be of type II. If the associated mapping becomes unstable by the dominant eigenvalue passing through -1, then the transition to chaos through intermittency is said to be of type III.

Type II intermittency requires consideration of at least two-dimensional maps, and consideration of such mappings and flows is deferred until Chapters 5 and 6.

The bifurcation associated with a type III intermittency transistion is considered in bifurcation theorem BF2. However, in contrast to the flip bifurcation leading to the period-doubling transistion to chaos, the flip bifurcation for type III intermittency must be subcritical, meaning that the nonlinear pieces of the map do not stabilize the mapping after the bifurcation, and the period 2 points are unstable rather than being stable. The growth in amplitude of the subharmonic frequency and the decay in amplitude of the fundamental, as though the system were proceeding to a period-doubling transistion, followed by a sudden chaotic burst are characteristic of type III intermittency. Furthermore, the probability distribution of the duration in time of the laminar phase has a long tail in contrast to the distribution to be expected from type I intermittency as sketched in Fig. 3.18. Further detail in regards to the intermittency transistion to chaos is contained in the excellent discussion in Bergé et al. (1984). Table 3.2 summarizes important properties of the intermittency transistions.

Table 3.2: Intermittency Types			
Type	Stability Loss	Archetype Function	Probability Distribution
I	$\lambda \to +1$	$f(x) = \mu + x + x^2$	Has sharp cutoff at $t \sim 1/\sqrt{\mu}$
II	$\lambda \to e^{\pm i\phi}$	Two-dimensional map (See Chapter 7)	
III	$\lambda \to -1$	$f(x) = \mu - x + x^2$	Has exponentially decaying tail (See Bergé et al., 1984)

3.6 Summary

In this chapter we have discussed universality for period-doubling and intermittency transistions to chaos. We considered in some detail the universal functions associated with these transistions, particularly for period doubling. From the universal functions for period doubling we obtained the universal scaling function for the distance between cycle elements. This scaling function was then used to predict scaling behavior for the Fourier amplitudes in a time evolving system undergoing period doubling as a system parameter is varied. The predictions of the period doubling route to chaos have been verified experimentally in a large number of physical systems. It is precisely this universality of chaotic behavior in very different physical systems that makes the study of transitions to chaos so interesting and full of promise. Universality gives theoretical predictions for diverse physical systems and depends only on the type of chaos exhibited.

Exercises

3.1 Consider the map $f_\mu(x) = x^2 - (\mu + 1)x$. Verify that it undergoes a flip bifurcation at $x = 0$, $\mu = 0$. Renormalize the map f_μ^2 following the outline sketched in Section 3.1 for the logistic map. Show that the same estimates are obtained for δ and μ_∞ but for a obtain $-2.239903\ldots$.

3.2 Show that the average duration of the laminar phase in the logistic map for μ near $\mu_c = 1 + 2\sqrt{2}$ scales as $1/\sqrt{|\mu - \mu_c|}$. Determine the shape of the probability distribution $P(t, \mu)$ and compare with Fig. 3.18.

3.3 Consider the function for type I intermittency to have the form

$$f(x) = x + nx^z + \ldots$$

near $x = 0$. This function clearly satisfies the boundary conditions (3.87). Find the auxiliary function $g_*(x)$ and the universal function $f_*(x)$ that is a fixed point of the doubling operator T defined in (3.85).

3.4 Consider a linearization of the doubling operator near

$$f_*(x) = x\left[1 - u(z - 1)x^{z-1}\right]^{\frac{-1}{(z-1)}}.$$

Write $f_\mu(x) = f_*(x) + \mu h_\lambda(x)$ where $h_\lambda(x)$ is considered to be an eigenfunction of the linear piece of T. That is to say

$$T[f_\mu(x)] = T[f_*(x) + \mu h_\lambda(x)] = f_*(x) + \lambda \mu h_\lambda(x) + \mathcal{O}(\mu^2).$$

Choose the auxiliary function $g_\mu(x)$ to be given by

$$g_\mu(x) = g_*(x) + \mu H_\lambda(x)$$

with the assumption that $H_\lambda(x) = x^{-p}$. Show that

$$h_\lambda(x) = \frac{1}{np}\left[x^{-p} - \left(x^{-(z-1)} - u(z-1)\right)^{p/(z-1)}\right] \Big/ \left[x^{-(z-1)} - u(z-1)\right]^{z/(z-1)},$$

where h_λ has been normalized so that its lowest order term $\simeq x^{2z-p-1}$. Show also that $\lambda = 2^{(p-z+1)/(z-1)}$. [Hint: Find the linear piece of the doubling operator as in (3.29) and the auxiliary function $H_\lambda(x)$ with its assumed form to solve the eigenvalue equation for $h_\lambda(x)$.]

CHAPTER FOUR

FRACTAL DIMENSION

A contemporary development with the emergence of Chaotic Dynamics has been the growing realization of the fractal nature of the world around us. The first and foremost reference on the nature of fractal geometry has to be Mandelbrot (1983). Mandelbrot was the first to point out and popularize the notion of fractal geometry as a useful and accurate representation for naturally occurring phenomena. The use of fractal geometry in the natural sciences is probably in its infancy despite a flood of papers describing applications in nearly all of them. These applications range from characterizing the convoluted surface of the brain (Majumdar and Prasad, 1988) to describing cracks (Mandelbrot et al.,1984). Two nontechnical articles by LaBrecque (1986, 1987)describe a large number of recent investigations focusing on fractal geometry in science. A recent resource letter in the *American Journal of Physics* (Hurd, 1988) provides a comprehensive source of references.

Despite the identification of fractals in nearly every branch of science, too frequently the recognition of fractal structure is not accompanied with any additional insight as to its cause. Often we do not even have the foggiest idea as to the underlying dynamics leading to the fractal structure. The chaotic dynamics of nonlinear systems, on the other hand, is one area where considerable progress has been made in understanding the connection with fractal sets. There is, however, always the difficulty that the nonlinear systems embodied in the mathematical models represent nature only imperfectly. Indeed, chaotic dynamics and fractal geometry have such a close relationship that one of the hallmarks of chaotic behavior has been the manifestation of fractal geometry, particularly for strange attractors in dissipative systems. For a practical definition we take a *strange attractor* to be an attracting set with fractal dimension.

Before giving precise definitions of fractal dimension, we first describe qualitative features characterizing fractal structures. The first property we mention is that of self-similarity. Upon repeated magnification fractal structures generally appear to look the same, which is to say that fractal structures generally have an inherent scale invariance. As one example we refer to Fig. 3.10. No matter what magnification, examination of the function obtained by taking the limit $n \to \infty$ of $F^{2^n}_{\mu_\infty}$ reveals battlements within battlements. A second example is the Z shaped figure following the chapter title. These examples in one and two dimensions illustrate clearly the phenomenon of self-similarity that generally accompanies fractal dimension.

The second feature of fractal structures is their lack of smoothness. Fractals always look jagged or disconnected. These features suggest the lack of differentiability that is part of fractal sets. The "spiked" nature of the curve in Fig. 3.10 is a reflection of this nondifferentiability.

Once sensitized to the possibilities of fractal geometry one no longer sees smooth surfaces and curves everywhere. As suggested by Mandelbrot (1983, p.1):

> *Clouds are not spheres, mountains are not circles, and bark is not smooth, nor does lightening travel in a straight line.*

4.1 Definitions of Fractal Dimension

Despite the fact that one important characterization of a self-similar curve or point set is its *fractal dimension*, there is not one unique meaning of fractal dimension. We focus on four different measures of fractal dimension that have been discussed in the literature. For further detail we refer the interested reader to Farmer et al. (1983b), Grassberger and Procaccia (1983b), and Farmer (1982). The recent work by Moon (1987, Chapter 6) considers several experimental examples and the problems of measuring fractal dimension. Beyond those presented in this chapter, subsequent chapters also contain applications and examples of computing fractal dimension.

The first definition we examine is called the *capacity dimension* and is denoted by d_{cap}. Let A denote the point set of interest, and assume A to be a bounded subset of Euclidean space $\mathbb{R}^m$. Let $N(\epsilon)$ denote the minimum number of m-dimensional cubes of side ϵ needed to cover A. As the size of ϵ diminishes, one expects the number $N(\epsilon)$ to increase. If for small ϵ, $N(\epsilon)$ increases according to the formula

$$N(\epsilon) \propto \epsilon^{-d_{\text{cap}}}, \tag{4.1}$$

then we say that d_{cap} is the capacity dimension of the set A. To be precise we define

$$d_{\text{cap}} = \lim_{\epsilon, \epsilon' \to 0} \frac{\log[N(\epsilon)/N(\epsilon')]}{\log[\epsilon'/\epsilon]}, \tag{4.2}$$

or

$$d_{\text{cap}} = \lim_{\epsilon \to 0} \frac{\log[N(\epsilon)]}{\log[1/\epsilon]}. \tag{4.3}$$

The capacity measure of the fractal dimension depends on the metric properties of the space $I\!R^m$ since the length ϵ is the crucial measure. This dimension d_{cap} also encompasses the familiar topological dimension: for a set A consisting of a single point, $N(\epsilon) = 1$ regardless of ϵ. For A a line of length L, $N(\epsilon) = L/\epsilon$ and for A a surface of area S, $N(\epsilon) = S/\epsilon^2$. In these cases we see immediately that $d_{\text{cap}} = 0, 1, 2$, respectively. The capacity dimension d_{cap} is often referred to as the Hausdorff dimension, although there is a technical difference that on most attractors is unimportant.

Intuitively, the dimension of a space or set denotes the amount of information necessary to specify a location. But a priori there is no reason why this information must be integer. Indeed, the second definition of fractal dimension that we use is called the *information dimension* and depends on the natural probability measure on the attracting set A. Let $\{C_i\}$ denote a covering of A by "cubes," C_i. Let P_i denote the fraction of images under the map that are in C_i, or for a continuous flow, the fraction of time spent in C_i. Let the ith cube C_i have side ϵ, and let $N(\epsilon)$ denote the total number of cubes in the covering $\{C_i\}$. Then the information dimension d_I is defined by

$$d_I = \lim_{\epsilon \to 0} \frac{I(\epsilon)}{\log(1/\epsilon)}, \tag{4.4}$$

where

$$I(\epsilon) = \sum_{i=1}^{N(\epsilon)} P_i \log(1/P_i). \tag{4.5}$$

In information theory $I(\epsilon)$ represents the amount of information necessary to specify the state of a system to within an accuracy ϵ. Equivalently, it is the amount of information gained by making a measurement that is uncertain by an amount ϵ.

In the capacity dimension d_{cap} each cube counts equally, regardless of the probability of there being points in the attractor in the given cube. The information dimension d_I takes into account the relative probability of the cubes in the covering $\{C_i\}$. Note that if the probability for each cube C_i is the same, then $I(\epsilon) = N(\epsilon)P \log[1/P]$ and $P = 1/N(\epsilon)$ giving $I(\epsilon) = \log[N(\epsilon)]$. Then according to (4.3) $d_I = d_{\text{cap}}$. For unequal probabilities $I(\epsilon) < \log[N(\epsilon)]$ and hence $d_I \leq d_{\text{cap}}$.

The third definition we give for fractal dimension is called the *correlation dimension*, and we denote it as d_{corr}. The basic idea behind this definition for fractal dimension is that the number of pairwise correlations in an ϵ "ball" about points on an attracting set scales as ϵ raised to some power. Specifically, we define

$$C(\epsilon) = \lim_{N \to \infty} \frac{1}{N^2} \sum_{\substack{i,j=1 \\ i \neq j}}^{N} \theta(\epsilon - \|\mathbf{x}_i - \mathbf{x}_j\|), \tag{4.6}$$

where $\theta(\alpha)$ is the Heaviside function, $\theta(\alpha) = 0(1)$ for $\alpha < 0(\alpha > 0)$. The notation $\|x_i - x_j\|$ denotes some convenient norm for the "distance" between the points x_i and x_j. As the value of ϵ decreases, clearly the number of pair correlations will decrease as well,

$$C(\epsilon) \propto \epsilon^{d_{\text{corr}}} \qquad (4.7)$$

and we have

$$d_{\text{corr}} = \lim_{\epsilon, \epsilon' \to 0} \frac{\log(C(\epsilon)/C(\epsilon'))}{\log(\epsilon/\epsilon')}. \qquad (4.8)$$

The correlation dimension also takes into account the population density on an attractor. Grassberger and Procaccia (1983b) show that the fractal dimensions discussed to this point satisfy the relation

$$d_{\text{corr}} \leq d_I \leq d_{\text{cap}} \qquad (4.9)$$

and that for many attractors of interest these dimensions are nearly equal (cf. Section 9.4).

The fourth definition for the fractal dimension is called the *Lyapunov dimension* d_L. Before giving the definition of this measure of dimension, we must generalize our definition of Lyapunov exponent given in Section 2.2 for one dimension to more than one dimension. In (2.10) it is the derivative of the iterated, one-dimensional map f that is relevant. For the m-dimensional map $\mathbf{f}(\mathbf{x})$ it is the corresponding object that is important, namely the Jacobian matrix $[\partial \mathbf{f}/\partial \mathbf{x}]$. The Jacobian matrix of the iterated map is given by

$$\left.\frac{\partial \mathbf{f}^n}{\partial \mathbf{x}}\right|_{\mathbf{x}_0} = \left.\frac{\partial \mathbf{f}}{\partial \mathbf{x}}\right|_{\mathbf{x}_{n-1}} \left.\frac{\partial \mathbf{f}}{\partial \mathbf{x}}\right|_{\mathbf{x}_{n-2}} \cdots \left.\frac{\partial \mathbf{f}}{\partial \mathbf{x}}\right|_{\mathbf{x}_0}, \qquad (4.10)$$

where the product of factors in (4.10) is matrix multiplication. Designate the magnitude of the eigenvalues of this matrix as $\mu_1(n) \geq \mu_2(n) \geq \cdots \geq \mu_m(n)$. The Lyapunov characteristic exponents are defined by

$$\lambda_i(\mathbf{x}_0) = \lim_{n \to \infty} \frac{1}{n} \ln[\mu_i(n)], \quad i = 1, 2, \ldots, m. \qquad (4.11)$$

As noted in (4.11), the Lyapunov exponents generally depend on the initial point. We assume on the basin of attraction for A that the characteristic exponents are the same for almost all points. Just as in one dimension, the characteristic exponents measure the average stretching along the characteristic directions of the Jacobian matrix. Section 9.2 contains a more detailed discussion of Lyapunov exponents in the multidimensional case and a consideration of their numerical computation.

The Lyapunov characteristic exponents have been made the basis of a measure of fractal dimension. For two exponents the Lyapunov dimension is defined as

$$d_L = 1 - \frac{\lambda_1}{\lambda_2}. \qquad (4.12)$$

For the case of m dimensions

$$d_L = k - \frac{\sum_{i=1}^{k} \lambda_i}{\lambda_{k+1}}, \tag{4.13}$$

where k labels the last λ_k for which $\lambda_1 + \lambda_2 + \cdots + \lambda_k \geq 0$. If $\lambda_1 < 0$ then define $d_L = 0$ and if $k = m$ then define $d_L = m$.

The relationship of this dimension in the general case to the other measures of fractal dimension still seems somewhat unsettled. Kaplan and Yorke (1978) seem to have been the first to conjecture that (4.13) is a meaningful measure of fractal dimension and for additional discussion see Frederickson et al. (1983) and Farmer (1982).

Several things are worth mentioning concerning the Lyapunov dimension d_L. First, in some simple cases d_L gives the same measure for fractal dimension as the other definitions. Second, the Lyapunov dimension is computationally much easier to obtain. Third, since most of the contribution to the calculation of Lyapunov exponents comes from those parts of an attractor that are visited most often, the Lyapunov dimension probably does not account for seldom visited regions as well as the other definitions of fractal dimension.

Table 4.1 summarizes the various definitions of fractal dimension.

Table 4.1: Definitions of Fractal Dimension			
Name	Symbol	Definition	Remarks
Capacity (Hausdorff)	d_{cap}	$d_{cap} = \lim_{\epsilon \to 0} \frac{\log N(\epsilon)}{\log(1/\epsilon)}$	Depends on metric properties of space only
Information	d_I	$d_I = \lim_{\epsilon \to 0} \frac{I(\epsilon)}{\log(1/\epsilon)}$ $I(\epsilon) = \sum_{i=1}^{N(\epsilon)} P_i \log(1/P_i)$	Depends on the frequency each cell is visited by the map
Correlation	d_{corr}	$d_{corr} = \lim_{\epsilon \to 0} \frac{\log C(\epsilon)}{\log \epsilon}$ $C(\epsilon) = \lim_{N \to \infty} \frac{1}{N^2} \sum_{i,j} \theta(\epsilon - \|\mathbf{x}_i - \mathbf{x}_j\|)$	Depends on the spatial correlations of points on the attractor
Lyapunov	d_L	$d_L = 1 - \lambda_1/\lambda_2$	See (4.12) for more than two dimensions

4.2 Classical Examples of Fractal Sets

Let us now consider some examples of self-similiar sets having fractal dimension. We first consider two standard examples: a classical Cantor set and the Koch snowflake curve.

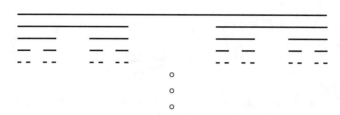

FIGURE 4.1   First four steps in the construction of the classical Cantor set.

The classical Cantor set is obtained by removing the middle third of the unit interval, followed by removal of the middle third of each remaining interval and so on ad infinitum. See Fig. 4.2. We compute the capacity dimension of this set by noting that at each step in the construction process the length ϵ is reduced by a factor of $\frac{1}{3}$ and the number of such lengths needed to cover the set is doubled. So for $\epsilon = (\frac{1}{3})^n$, $N(\epsilon) = 2^n$. Consequently

$$d_{\text{cap}} = \lim_{n \to \infty} \frac{\log(2^n)}{\log(3^n)} = \frac{\log 2}{\log 3} = 0.6309\ldots.$$

Consider now the Koch snowflake, whose construction is indicated in Fig. 4.2. The sides of an equilateral triangle are each divided in thirds. Each of the middle thirds is made the base of an equilateral triangle protruding outward from the original triangle. In this case the length ϵ is again reduced by a factor of $\frac{1}{3}$ each time. The number of lengths needed clearly increases by a factor of 4. So for $\epsilon = (\frac{1}{3})^n$, $N(\epsilon) = 3 \cdot 4^n$, and we obtain for the capacity dimension

$$d_{\text{cap}} = \lim_{n \to \infty} \frac{\log 3 + n \log 4}{n \log 3} = \frac{\log 4}{\log 3} = 1.2618\ldots.$$

The unfamiliarity of such sets is further emphasized when we note that the length or measure of the Cantor set is zero and yet it has nonzero fractal dimension. We are used to having sets with zero length, such as points, to also have zero dimension. Similarly, the Koch snowflake has infinite length and yet encloses finite area. Self-similar sets such as these can be wonderfully strange.

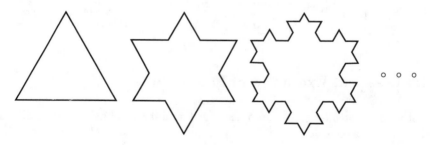

FIGURE 4.2   Construction of the Koch snowflake.

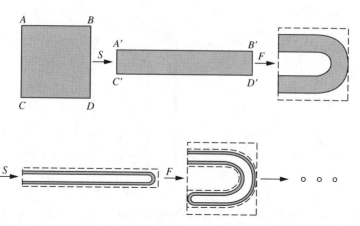

FIGURE 4.3   Sequence of operations on the unit square to produce the horseshoe attractor.

As a somewhat more complicated example of an object with fractal dimension we consider the horseshoe attractor. As we have already learned the basic operations for creating an attractor consist of stretching and folding. Figure 4.3 shows a sequence of operations performed on the unit square. The first operation (S) consists of stretching by a factor of 2 in the x direction and compressing by a factor of 2α in the y direction. If $\alpha > 1$, then this transformation contracts areas as is the case for dissipative systems. The second operation (F) is then to fold the rectangle $A'B'C'D'$ and fit it back into the square $ABCD$. The shape of this folded rectangle is the source of the name for this attractor. Then we repeat this sequence $SFSF\ldots$ an infinite number of times. more precisely, let $s = F \circ S$ denote the composite map on the unit rectangle R, $s : R \rightarrow R$. The attracting set A is defined as the intersection of all the sets formed by iterations of s on R, i.e.,

$$A = \cap_{n=0}^{\infty} s^n(R). \tag{4.14}$$

The fact that A is an attractor is guaranteed by the value of $\alpha > 1$, which ensures that the image of every point in R under s is inside R.

In order to compute d_L we assume that α is sufficiently large so that the effect of the bend can be ignored. Note that nearby initial points have exponential separation because of stretching in x. Indeed we can easily obtain the Lyapunov characteristic exponents for almost all points. In the x direction the distances increase by the factor 2 under each iteration, and thus $\lambda_1 = \ln 2$, and similarly in the y direction $\lambda_2 = -\ln 2\alpha$.

$$\lambda_1 + \lambda_2 = \ln 2 - \ln 2 - \ln \alpha = -\ln \alpha < 0.$$

So in applying the definition (4.13), k is clearly equal to 1 and

$$d_L = 1 + \frac{\ln 2}{\ln 2\alpha}.$$

For $\alpha > 1$ we obtain $1 < d_L < 2$.

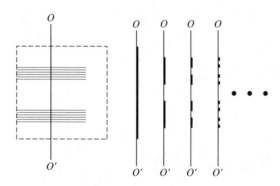

FIGURE 4.4   Images of the unit square R on the section $O-O'$ under successive iterates of the stretch-fold map for the horseshoe attractor.

We can obtain a measure of the fractal dimension of A in another way by using the capacity dimension. Consider a vertical section through the unit square as sketched in Fig. 4.4 and the resulting images of $s^n(R)$ on this section for increasing values of n. We know that the size of each intersection is reduced by the factor $1/2\alpha$ under each iteration of s. Thus the size of ϵ in the nth iteration is given by $1/(2\alpha)^n$. Also it is clear that the number of such intervals needed to cover the set of intersection points doubles, and so in the nth iteration $N(\epsilon) = 2^n$. The structure of this set on the section line is very similar to the construction of the classical Cantor set. Adding 1 to (4.3) for the dimension perpendicular to our line of section, we obtain

$$d_{\text{cap}} = 1 + \lim_{n\to\infty} \frac{\log 2^n}{\log(2\alpha)^n} = 1 + \frac{\log 2}{\log 2\alpha},$$

which is identical to the Lyapunov dimension obtained for A.

In addition to its value as an instructive example, the essential features of the horseshoe attractor are found in many systems observed to produce strange attractors.

4.3 Attractor for the Universal Quadratic Function

Moving to a less contrived example, let us consider the attractor set of a 2^n-supercycle in the limit that $n \to \infty$. Let ϵ be the maximum length to cover the 2^n-supercycle with exactly one point in the supercycle in each segment. Then $N(\epsilon)$ is, of course, just 2^n. Now step up to a 2^{n+1}-supercycle. We know that the elements of the supercycle get closer together in going from 2^n to 2^{n+1}. The factor by which they get closer together is given by the function σ of Fig. 3.11 in the limit that $n \to \infty$. From this figure we see that half the elements get closer by approximately $1/\alpha^2$ and half by approximately $1/\alpha$. Those elements represented by rationals between $(0, \frac{1}{4})$ get closer by approximately $1/\alpha^2$, and those elements represented by rationals in the interval

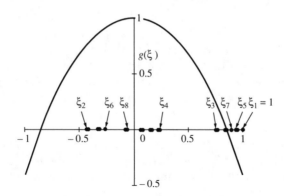

FIGURE 4.5   The universal function $g(\xi)$ near $\xi = 0$ with the first few cycle elements marked.

$(\frac{1}{4}, \frac{1}{2})$ get closer by approximately the factor $1/\alpha$. Following Schuster (1984), we find that the new ϵ', in going from a 2^n-supercycle to a 2^{n+1}-supercycle, is approximately given by the average of ϵ scaled by $1/\alpha$ half the time and by $1/\alpha^2$ the other half. That is

$$\epsilon' \simeq \left(2^n\frac{\epsilon}{\alpha} + 2^n\frac{\epsilon}{\alpha^2}\right)/2^{n+1} = \frac{\epsilon}{2}\left(\frac{1}{\alpha} + \frac{1}{\alpha^2}\right).$$

Now calculate the capacity dimension according to (4.2).

$$d_{\text{cap}} = \lim_{\epsilon \to 0} \frac{\log[2^{n+1}/2^n]}{\log[\epsilon/\frac{\epsilon}{2}(\frac{1}{\alpha} + \frac{1}{\alpha^2})]} = -\frac{\log 2}{\log[\frac{1}{2}(\frac{1}{\alpha} + \frac{1}{\alpha^2})]} = 0.54387\ldots \; .$$

This crude estimate for the fractal dimension of the attractor for $F_\mu^{2^n}$ with $\mu = \mu_\infty$ and $n \to \infty$ was obtained without much effort, but it is not clear exactly what one should do to obtain a more accurate value. With this object in mind we consider the more rigorous procedure of Grassberger (1981) for computing d_{cap}.

As noted earlier the universal function $g(\xi)$ encodes near its maximum a complete set of cycle elements (see Section 3.1). This set of cycle elements will have the same fractal dimension as the set of all cycle elements for $F_{\mu_n}^{2^n}$, $n \to \infty$. Figure 4.5 shows this function $g(\xi)$ for $\xi \in [-1, 1]$ and the first few cycle elements have also been marked and labeled. Because of the normalization of (3.12) it is clear that

$$\xi_0 = 0, \; \xi_1 = g(0) = 1, \; \xi_2 = g(\xi_1) = g(1) = -1/\alpha, \; \xi_3 = g(1/\alpha).$$

Furthermore we see that the whole attractor lies in the interval $[\xi_2, \xi_1]$. It requires only a simple induction argument to show that

$$\xi_{2k} = -\left(\frac{1}{\alpha}\right)\xi_k, \quad k = 1, 2, 3, \ldots \; . \tag{4.15}$$

Using (4.15) we find that the distance between ξ_2 and ξ_4 is $1/\alpha$ times the distance between ξ_2 and ξ_1. Indeed the set $\{\xi_k, k \text{ even}\}$ lies in $[\xi_2, \xi_4]$ and is exactly similar to the whole set $\{\xi_k\}$ but scaled by a factor $(-1/\alpha)$. The set $\{\xi_k, k \text{ odd}\}$ lies in the disjoint interval $[\xi_3, \xi_1]$. Because $g(\xi)$ is monotonic on $[\xi_3, \xi_1]$ and smooth, and since $\{\xi_k, k \text{ even}\}$ is obtained from $\{\xi_k, k \text{ odd}\}$ by application of g, the distribution of points in $[\xi_3, \xi_1]$ is qualitatively similar to the pattern of points in $[\xi_2, \xi_4]$ but somewhat distorted by the changing slope of $g(\xi)$ in this interval.

The set of cycle elements $\{\xi_k\}$ separates into two subsets each of which is similar to it. The same can be said of each of these subsets ad infinitum. This is, of course, the meaning of self-similarity for this point set of 2^n-supercycle elements as $n \to \infty$.

Let us now cover the intervals by small intervals of length ϵ. We denote by $N_{[2,4]}(\epsilon)$ the number of lengths ϵ needed to cover the interval $[\xi_2, \xi_4]$, with the extension of this notation to other intervals being obvious. From our earlier remark relating the length of the interval $[\xi_2, \xi_4]$ and $[\xi_2, \xi_1]$, it is clear that

$$N_{[2,4]}(\epsilon) = N(\alpha\epsilon). \tag{4.16}$$

Furthermore,

$$N(\epsilon) = N_{[2,4]}(\epsilon) + N_{[3,1]}(\epsilon). \tag{4.17}$$

Since $g''(\xi) < 0$ in $[\xi_3, \xi_1]$, $|g'(\xi)|$ decreases as one goes from ξ_3 to ξ_1. Thus $N_{[3,1]}(\epsilon)$ intervals of length $|g'(\xi_3)|\epsilon$ are not enough to cover $[\xi_2, \xi_4]$. Similarly, $N_{[3,1]}(\epsilon)$ intervals of length $|g'(\xi_1)|\epsilon$ is more than enough. As a consequence we have the inequalities

$$N_{[2,4]}\big(|g'(\xi_1)|\epsilon\big) < N_{[3,1]}(\epsilon) < N_{[2,4]}\big(|g'(\xi_3)|\epsilon\big). \tag{4.18}$$

The fractal dimension of each subinterval will be the same because of the self-similarity. This allows us to relate the constant implicit in (4.1) for each subinterval using (4.16) and (4.17). Subscripting the constants by intervals we have

$$C_{[2,4]} = C/\alpha^{d\text{cap}} \qquad \text{and} \qquad C_{[3,2]} = C\left(1 - \frac{1}{\alpha^{d\text{cap}}}\right).$$

Substituting these results into (4.18) we obtain

$$\frac{1}{\alpha^{d\text{cap}}} + \frac{1}{(\alpha|g'(\xi_1)|)^{d\text{cap}}} < 1 < \frac{1}{\alpha^{d\text{cap}}} + \frac{1}{(\alpha|g'(\xi_3)|)^{d\text{cap}}}. \tag{4.19}$$

Differentiating (3.11) twice and evaluating at $\xi = 0$ shows that $g'(1) = -\alpha$ and from the representation (3.14) with values given in the subsequent table we find $g'(\xi_3) = -2.12703$. Substituting into (4.19) we find the bounds

$$0.5245\cdots < d_{\text{cap}} < 0.5519\cdots$$

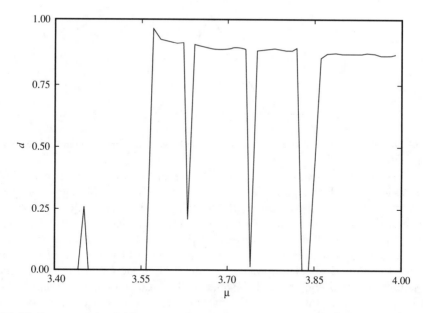

FIGURE 4.6   Fractal dimension of the logistic map attractor as a function of the map parameter μ.

Successively tighter bounds are achieved by dividing the interval $[\xi_3, \xi_1]$ into 2^n subintervals and using the approximate similarity of each of them to a subinterval of $[\xi_2, \xi_4]$. In this way Grassberger achieves the bounds

$$0.53763 < d_{\text{cap}} < 0.53854.$$

Note that the estimate obtained at the beginning of this section lies outside this range.

4.4 Summary

To some extent the examples presented here for fractal dimension are special. For generic dynamical systems one cannot use analytic methods to calculate a fractal dimension. Numerical methods are a necessity, but one does not usually obtain satisfactory results by a naive application of the formulas in this section. For example, a computation of the capacity dimension d_{cap} requires an adequately small ϵ so that one obtains reasonable convergence for the $\epsilon \to 0$ limit. However, small ϵ requires a large number of points on the attractor and storage requirements can rapidly become excessive requiring special considerations. Section 7.4 discusses the Hènon map and Fig. 7.15 shows this convergence. These issues are discussed in detail by Grassberger and Procaccia (1983a) wherein certain subtleties associated with computations of correlation dimension are also discussed. Refer also to Grassberger (1983) for a specific consideration of the Hènon map.

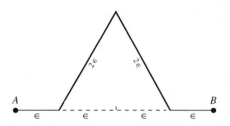

FIGURE 4.7   Interval construction for the modified Koch curve used in Exercise 4.2.

Figure 4.6 is an example of results obtained by calculation of correlation dimension. This figure plots the correlation dimension of the logistic map as a function of the map parameter μ. The computation is crude but illustrates some of the problems encountered in such computations. The fractal dimension of a periodic orbit is of course zero. If, however, the convergence to this orbit is weak, then it requires a very large number of iterations of the map before the numerical calculation of the fractal dimension reflects the fact that it is indeed a periodic orbit. The spikes in Fig. 4.6 represent a computation of d_{corr} at a value for μ where the number of iterations of the map was insufficient to really accurately sample the nature of the attractor. A second problem evident in Fig. 4.6 is the rather large increment in μ. Figure 4.6 shows the fractal dimension is zero in those regions along the μ-axis corresponding to periodic orbits. Figure 4.6 should be compared with Fig. 2.8.

Irrespective of computational difficulties that can in part be overcome, there exists a more fundamental problem in deciding what is to be learned from a fractal dimension. A fractal dimension by itself does not seem to be a very discriminating characterization of strange sets. Possibly a whole sequence of fractal dimensions, as discussed in Section 9.4, are needed to adequately characterize strange sets. Possibly it is the singularities of distribution functions or measures on fractal sets that provide a more discriminatory measure. Such topics are beyond the scope of this introductory work, and we refer the interested reader to the research literature, which includes Amritkar et al. (1987), Sarkar (1987), and Halsey et al. (1986).

Exercises

4.1 Show that the length of the classical Cantor set is 0 and that the length of the Koch snowflake is unbounded. Find the area of a Koch snowflake that starts from a triangle with sides of length 1.

4.2 Consider interval subdivisions that proceed according to Fig. 4.7. Each interval is subdivided in four pieces, and the two middle quarters form the base of an equilateral triangle. Continue this construction n times as $n \to \infty$, and show that the fractal dimension of the curve obtained between A and B is 1.292... and that it has infinite length.

4.3 As an example for comparing fractal dimensions and verifying (4.9) in one specific case, consider once again the construction for the classical Cantor set. Establish a probability measure on this set in the following way. The probability on each remaining subinterval after deletion is reduced by the same factor as the length of the interval, except that the right-hand subinterval gets the measure from the omitted middle third. For example, after the first deletion, the remaining third on the right has probability $\frac{2}{3}$ and the third on the left has probability $\frac{1}{3}$. After the next deletion the intervals from left to right have probabilities of $\frac{1}{9}, \frac{2}{9}, \frac{2}{9}, \frac{4}{9}$, etc. Using the fact that the pairwise correlation is proportional to $\sum_i P_i^2$, compute d_I and d_{corr}. Sketch the probability distribution.

CHAPTER FIVE

DIFFERENTIAL DYNAMICS

Who can understand his errors?
(Psalms 19:12)

In actual fact it is a rarity that a dynamical system can be described by mappings. The far more common case is that differential equations are required, perhaps even partial differential equations. In this chapter we focus on dynamical systems whose evolution is modeled by ordinary differential equations. These remarks are not meant to imply that the results in the foregoing chapters on the period-doubling or intermittency cascades into chaos are not applicable. On the contrary, a large number of systems that are properly modeled by differential equations are such that the dissipation in the system effectively makes it one-dimensional (see Section 3.4). There is, however, a rich variety of alternative paths by which a system may become chaotic. Consequently, we turn to a description in terms of differential equations after having established much of the appropriate language and concepts by looking at one-dimensional maps.

In this chapter we remind the reader of many basic results in the theory of ordinary differential equations. We apply these results to understanding systems exhibiting chaotic behavior.

In order to make our remarks specific and concrete it is often helpful to have a familiar system before us on which we can "demonstrate" the concepts. For this purpose we choose a driven, damped pendulum. With an appropriate scaling of variables, Newtonian dynamics gives the following equation of a simple pendulum, driven by a periodic force and subject to damping proportional to the velocity:

$$\ddot{x} + \Omega^2 \sin x = \epsilon(-\alpha\dot{x} + f\cos\omega t). \tag{5.1}$$

The quantity ϵ serves the purpose of tagging those quantities that we wish to consider as small, in this case α and f. In any subsequent analysis ϵ may be set equal to 1 after all ordering of small quantities has been completed, assuming that α and f are suitably small.

Equation (5.1) is of the general type

$$\dot{\mathbf{x}} = \mathbf{f}_\mu(\mathbf{x}), \tag{5.2}$$

where μ is a map parameter and $\mathbf{x} \in \mathbb{R}^n$. In the case of (5.1) the parameter μ might be identified with α, f, or ω depending on whether the degree of damping, the amplitude of the drive, or the frequency of the drive is of interest. Equation (5.1) can be written in the form (5.2) in the usual way by defining a new variable $y = \dot{x}$. To obtain the autonomous form, we let $t \rightarrow \theta$ and then add $d\theta/dt = 1$ to the differential system.

We refer to a solution of (5.2) as a *trajectory* and note that the trajectory depends on the given initial conditions. The basic theorem for the existence and uniqueness of solutions to differential equation systems such as (5.2) is given in all textbooks on the theory of ordinary differential equations. We will always restrict our discussion to systems for which the existence of solutions is assured. We often refer to the trajectories resulting from a neighborhood of initial conditions as a *flow*.

5.1 Linearization

The dynamics determined by the vector field $\mathbf{f}_\mu(\mathbf{x})$ on the right-hand side of (5.2) can be highly nonlinear and complicated. However, as with maps, attracting sets and fixed points are of special interest. The fixed points are readily recognized as just those values of $\mathbf{x}$ for which $\mathbf{f}_\mu(\mathbf{x}) = 0$. Fixed points can, of course, be stable or unstable, and so the behavior of trajectories in the neighborhood of fixed points is significant. To study this behavior we must linearize (5.2) in the neighborhood of such a fixed point.

Suppose $\mathbf{x}_*$ to be a fixed point of (5.2), i.e., $\mathbf{f}(\mathbf{x}_*) = 0$, where for the moment we suppress the map parameter μ. We now make a Taylor expansion of $\mathbf{f}(\mathbf{x})$ around $\mathbf{x}_*$, assuming of course that $\mathbf{f}$ has adequate differentiability properties.

$$\mathbf{f}(\mathbf{x}) = \mathbf{f}(\mathbf{x}_*) + D\mathbf{f}(\mathbf{x}_*)(\mathbf{x} - \mathbf{x}_*) + \cdots, \tag{5.3}$$

where $D\mathbf{f}(\mathbf{x}_*)$ is the matrix of functions $\partial f_i/\partial x_j$ evaluated at the fixed point $\mathbf{x}_*$. Since $\mathbf{f}(\mathbf{x}_*) = 0$, for small $(\mathbf{x} - \mathbf{x}_*)$ the dynamics is determined solely by the linear term.

We let $\mathbf{y} = \mathbf{x} - \mathbf{x}_*$, $D\mathbf{f}(\mathbf{x}_*) = A$, and study the linear system

$$\dot{\mathbf{y}} = A\mathbf{y}. \tag{5.4}$$

As indicated earlier there is no loss of generality in focusing on the autonomous case since we can always expand the system to make it autonomous.

Equation (5.4) is regarded as a valid approximation to (5.2) only in a sufficiently small neighborhood of the fixed point $\mathbf{x}_*$. A specific solution to (5.4) with initial condition $\mathbf{y}_0$ is given by

$$\mathbf{y}(t) = e^{tA}\mathbf{y}_0, \tag{5.5}$$

as is easily verified by substitution. However, being interested in qualitative information (stability, asymptotic behavior, etc.), we are more interested in the general solution to (5.4) obtained by the linear superposition of n linearly independent solutions $\mathbf{y}_i(t), i = 1, \ldots, n$.

$$\mathbf{y}(t) = \sum_{i=1}^{n} c^i \mathbf{y}_i(t), \tag{5.6}$$

where the n unknown constants are determined by the initial conditions. If the matrix A has n distinct eigenvalues $\lambda_i, i = 1, \ldots, n$, then the linearly independent solution vectors can be chosen to be

$$\mathbf{y}_i(t) = e^{t\lambda_i} \mathbf{v}_i. \tag{5.7}$$

If the eigenvalue λ_i turns out to be complex, then both λ_i and λ_i^* are eigenvalues with $\mathbf{v}_i$ and $\mathbf{v}_i^*$ as eigenvectors. Real, linearly independent solution vectors can be obtained by taking the real and imaginary parts of (5.7) in the case of complex eigenvalues and vectors.

For a concrete example of these ideas consider system (5.1) with $f = 0$ so that the system, has a fixed point. Written as a first-order system, (5.1) becomes ($\epsilon = 1$)

$$\dot{x} = y,$$
$$\dot{y} = -a^2 \sin x - \alpha y. \tag{5.8}$$

The system (5.8) clearly has a fixed point at $(x, y) = (0, 0)$. The matrix Df at this fixed point is given by

$$\begin{pmatrix} 0 & 1 \\ -a^2 & -\alpha \end{pmatrix}$$

with corresponding linearly independent solutions given by

$$\mathbf{y}_1(t) = e^{-\alpha t/2} \left[\begin{pmatrix} 1 \\ -\alpha/2 \end{pmatrix} \cos \beta t - \begin{pmatrix} 0 \\ \beta \end{pmatrix} \sin \beta t \right],$$
$$\mathbf{y}_2(t) = e^{-\alpha t/2} \left[\begin{pmatrix} 1 \\ -\alpha/2 \end{pmatrix} \sin \beta t + \begin{pmatrix} 0 \\ \beta \end{pmatrix} \cos \beta t \right], \tag{5.9}$$

where $\beta = a\sqrt{1 - \alpha^2/4a^2}$.

If the system has degeneracies so that λ is an eigenvalue of A with multiplicity k, then we must compute the generalized eigenvectors. This procedure is fully discussed in Braun (1983). We suppose here that $k = 2$ and solve for $\mathbf{v}$ and $\mathbf{u}$, where

$$A\mathbf{v} = \lambda\mathbf{v}, \quad (A - \lambda I)\mathbf{u} \neq 0, \quad (A - \lambda I)^2 \mathbf{u} = 0, \tag{5.10}$$

giving the independent solution vectors

$$\mathbf{y}_1(t) = e^{\lambda t} \mathbf{v},$$
$$\mathbf{y}_2(t) = e^{\lambda t} [\mathbf{u} + t(A - \lambda I)\mathbf{u}]. \tag{5.11}$$

In any case we find n linearly independent solutions $\mathbf{y}_i(t), i = 1, \ldots, n$ forming a basis for the space of solutions of (5.4). From these n solutions we form the *fundamental solution matrix*

$$Y(t) = [\mathbf{y}_1(t), \mathbf{y}_2(t), \ldots, \mathbf{y}_n(t)] \tag{5.12}$$

having the linearly independent solution vectors $\mathbf{y}_i(t), i = 1, \ldots, n$ as its columns. If $\mathbf{y}_0$ is an arbitrary initial point, then it is easy to show that $Y(t)Y(0)^{-1}\mathbf{y}_0$ is a solution of (5.4), and consequently by the uniqueness of solutions we have that

$$e^{tA} = Y(t)Y(0)^{-1}. \tag{5.13}$$

If $\mathbf{y}_0(s)$ denotes a family of initial conditions, then the flow $\mathbf{y}(t; s)$ is given by $Y(t)Y(0)^{-1}\mathbf{y}_0(s)$.

We may regard $e^{tA} = Y(t)Y(0)^{-1}$ as a mapping of $I\!R^n$ to $I\!R^n$ and, as we might expect, the linear subspaces generated by the eigenvectors of A have some special properties under this map. If $\mathbf{v}$ is an eigenvector of A with eigenvalue λ, then

$$e^{tA}\mathbf{v} = \sum_{j=0}^{\infty} \frac{(tA)^j}{j!}\mathbf{v} = \sum_{j=0}^{\infty} \frac{(t\lambda)^j}{j!}\mathbf{v} = e^{t\lambda}\mathbf{v}. \tag{5.14}$$

We see that the flow map e^{tA} applied to $\mathbf{v}$ gives simply a scalar times $\mathbf{v}$, i.e., the linear subspace spanned by $\mathbf{v}$ is invariant under e^{tA}. Here we have tacitly assumed that the eigenvalue λ was real; if λ is complex, then from (5.14) we see that it is the space spanned by $\{\mathrm{Re}(\mathbf{v}), \mathrm{Im}(\mathbf{v})\}$, which is invariant under e^{tA}. In short the eigenspaces of A are invariant under the flow map e^{tA}.

It is convenient to divide these invariant subspaces into three groups according to whether $\mathrm{Re}(\lambda_i)$ is less than, equal to, or greater than zero. From these three groups we form invariant subspaces of $I\!R^n$ called the *stable manifold*, the *center manifold*, and the *unstable manifold*. Specifically,

E_s is the subspace of $I\!R^n$ spanned by the eigenvectors of A $\{\mathbf{v}_i\}$ such that $\mathrm{Re}(\lambda_i) < 0$.

E_c is the subspace of $I\!R^n$ spanned by the eigenvectors of A $\{\mathbf{v}_i\}$ such that $\mathrm{Re}(\lambda_i) = 0$.

E_u is the subspace of $I\!R^n$ spanned by the eigenvectors of A $\{\mathbf{v}_i\}$ such that $\mathrm{Re}(\lambda_i) > 0$.

From (5.14) it is clear that solutions in E_s are characterized by exponential decay, those in E_u by exponential growth, and those in E_c by neither exponential decay nor growth.

As an example consider the matrix

$$\begin{pmatrix} 1 & 2 & 0 \\ 1 & 0 & 0 \\ 0 & 0 & 0 \end{pmatrix} \tag{5.15}$$

with eigenvalues $\lambda = 0, -1, 2$ and corresponding eigenvectors $(0,0,1), (1,-1,0),$ $(2,1,0)$. The invariant manifolds are sketched in Fig. 5.1.

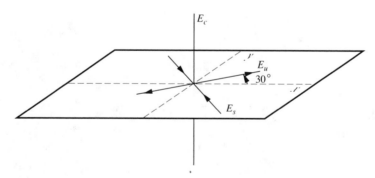

FIGURE 5.1   Sketch of the invariant manifolds for the matrix in (5.15).

5.2 Invariant Manifolds

As we turn our attention to the nonlinear aspects of our system, we note that nothing beyond the most general statements can be said about the solutions to (5.2). It is almost never possible to write down a solution in closed form, and many different techniques including averaging, asymptotic analysis, numerical integration, etc., should be applied in order to gain as much qualitative information about the solutions as possible.

A fixed point $\mathbf{x}_*$ is called *hyperbolic* if $D\mathbf{f}(\mathbf{x}_*)$ has no zero or purely imaginary eigenvalues, i.e., if $D\mathbf{f}(\mathbf{x}_*)$ has no center manifold. A hyperbolic fixed point has "nice" properties in that for some neighborhood $U \subset \mathbb{R}^n$ of the fixed point $\mathbf{x}_*$, there is a homeomorphism that maps the trajectories in the nonlinear flow of (5.2) to the trajectories in the linear flow generated by $D\mathbf{f}(\mathbf{x}_*)$. The homeomorphism preserves the sense of the flow and may be chosen to even preserve the parameterization in t. We are never interested in actually finding this homeomorphism but are only interested in knowing of its existence so that information gained about the linear flow may be applied with confidence to the nonlinear flow. A proof of these statements may be found in Irwin (1980).

Even more can be said: The invariant manifolds of the linear flow in the neighborhood of the fixed point $\mathbf{x}_*$ are tangent to corresponding manifolds in the nonlinear flow. These manifolds may only exist locally but satisfy the celebrated Center Manifold theorem of dynamics.

Center Manifold *Let $\mathbf{f}_\mu(\mathbf{x})$ be a C^r vector field on $\mathbb{R}^n$ that satisfies $\mathbf{f}_\mu(\mathbf{x}_*) = 0$. The spectrum of $D\mathbf{f}_\mu(\mathbf{x}_*)$ is divided into three*

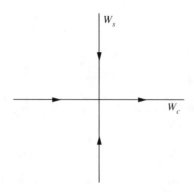

FIGURE 5.2   Invariant manifolds at (0,0) for the system of Equation (5.16).

sets σ_s, σ_c, and σ_u where

$$\lambda \in \begin{cases} \sigma_s, & \text{Re}(\lambda) < 0, \\ \sigma_c, & \text{Re}(\lambda) = 0, \\ \sigma_u, & \text{Re}(\lambda) > 0. \end{cases}$$

The generalized eigenspaces of σ_s, σ_c, and σ_u are E_s, E_c, and E_u, respectively. Then there exist C^r stable and unstable invariant manifolds W_s and W_u tangent to E_s and E_u at $\mathbf{x}_*$ and a C^{r-1} center manifold W_c tangent to E_c at $\mathbf{x}_*$. The manifolds W_s, W_u, and W_c are invariant under the flow generated by $\mathbf{f}_\mu$. The manifolds W_s and W_u are unique but W_c need not be.

Although we will frequently invoke the Center Manifold theorem and return subsequently to study various aspects of its implications more thoroughly, it is best here to examine two simple examples.

Consider the dynamical system

$$\dot{x} = xy + x^2,$$
$$\dot{y} = -y - x^2 y. \tag{5.16}$$

One readily finds the only fixed point of this system to be at $(x, y) = (0, 0)$, and the eigenvalues of $D\mathbf{f}(0)$ to be -1 and 0 with corresponding eigenvectors $\hat{\mathbf{y}}$ and $\hat{\mathbf{x}}$ denoting unit vectors along the y and x-axis, respectively. We find then $E_s = \text{span}\{\hat{\mathbf{y}}\}$ and $E_c = \text{span}\{\hat{\mathbf{x}}\}$. From (5.16) directly we see that the x-axis is an invariant set under the flow and forms the center manifold W_c, tangent to (in this case identical to) the center eigenspace E_c. See Fig. 5.2 for a sketch of these manifolds.

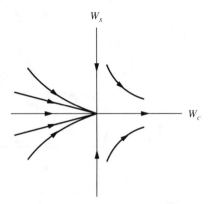

FIGURE 5.3   Solution curves of (5.17) showing the nonuniqueness of the center manifold.

The direction of the flow on W_c for the system (5.16) cannot be determined from linearized equations. On W_c we find $x(t) = x_0/(1 - tx_0)$ giving $x(t) \to 0$ as $t \to \infty$ for $x_0 < 0$. For $x_0 > 0$ we find $x(t)$ becomes unbounded in finite time.

A simplification of system (5.16) to

$$\dot{x} = x^2,$$
$$\dot{y} = -y, \tag{5.17}$$

can illustrate some of the possible complications associated with the center manifold W_c. The parameterized solutions of (5.17) are $x(t) = x_0/(1 - tx_0)$, $y(t) = y_0 e^{-t}$. Elimination of t gives $y(x) = (y_0/e^{1/x_0})e^{1/x}$, which constitute the solution curves of (5.17). A qualitative sketch of these curves is given in Fig. 5.3.

All of the solution curves for $x < 0$ come into 0 with $dy/dx = 0$, but only the x-axis does so for $x > 0$. Thus any of the curves on the left, since each joins with the positive x-axis, can be the center manifold W_c. The choice of the entire x-axis as W_c is the only analytic one at $(x, y) = (0, 0)$.

We summarize the general situation by a sketch suggesting the relationship between the linear eigenspaces and the invariant manifolds of the nonlinear system (5.2) in Fig. 5.4. Note once again that the flow direction(s) in W_c cannot be determined from the linearization.

As a further example consider system (5.1) with $\epsilon = 0$, i.e., a simple pendulum. We find the fixed points $(x, y) = (0, 0)$, $(n\pi, 0)$, $n = \pm1, \pm2, \ldots$, with the corresponding eigenvalues for the linearized system $\lambda = \pm ia$, $\pm a$. Consequently, the phase curves near the origin are closed and encircle it. This type of fixed point is called a *center*. The fixed points at multiples of π are referred to as saddles since they have a stable and an unstable manifold. Refer to Fig. 5.5 for a sketch.

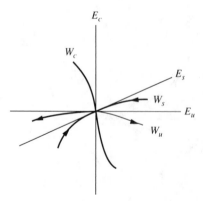

FIGURE 5.4   Relationship between invariant manifolds to those of the linearization.

The basic existence and uniqueness theorem for differential equations ensures that the stable manifolds of two distinct fixed points cannot intersect. Likewise, the unstable manifolds of two distinct fixed points cannot intersect. Neither can W_s or W_u intersect itself. However, intersections of W_s and W_u for distinct or even the same fixed point are possible. In Fig. 5.5 a portion of the stable and unstable manifolds for the two fixed points at either side of the center are identical but with opposite stability properties.

5.3 The Poincaré Map

We know that attracting sets need not be fixed points. Systems undergoing forced oscillations very often settle into an attracting periodic orbit in phase space called a *limit cycle*. An example of such a system is the driven, damped pendulum of Eq. (5.1). The attractor set is not the same for all choices of the

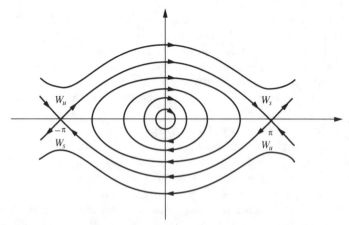

FIGURE 5.5   Phase-space trajectories for a simple pendulum with the invariant manifolds identified.

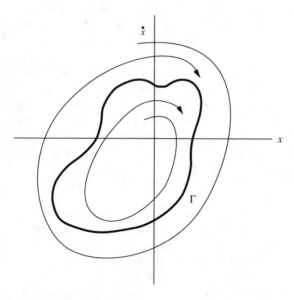

FIGURE 5.6   Sketch of a stable limit cycle Γ for a forced, nonlinear oscillator.

parameters α, f, and ω. However, for $\omega/\Omega \simeq 10$ and $\alpha/\Omega \sim f/\Omega \sim 0.25$ all phase-space trajectories are attracted to a limit cycle in phase space. Such a limit cycle Γ is pictured qualitatively in Fig. 5.6, but not all limit cycles are stable as the one sketched. Unstable limit cycles can also occur.

Sometimes one can solve for the periodic orbit and linearize about it to study system properties. More often, however, the flow is studied numerically by consideration of Poincaré sections. Let Γ denote a periodic trajectory in $I\!R^n$ and denote the flow map generated by the vector field $\mathbf{f}_\mu(\mathbf{x})$ of (5.2) as $\phi_t(\mathbf{x})$. A *Poincaré section* for the flow ϕ_t in a neighborhood of Γ is an $(n-1)$-dimensional hypersurface Σ *transverse* to the flow. To be transverse means

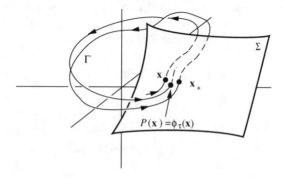

FIGURE 5.7   Sketch of a periodic orbit Γ and a nearby trajectory as they intersect a Poincaré section Σ.

that the normal to the surface Σ, $\hat{n}(x)$, is nowhere orthogonal to $f_\mu(x)$ for $x \in \Sigma$, i.e., $f_\mu(x) \cdot \hat{n}(x) \neq 0$ for all $x \in \Sigma \subset \mathbb{R}^n$. We also require Σ to be such that there is a single intersection of Γ with Σ. Refer to Fig. 5.7 for a sketch of the situation.

The Poincaré map or first return map is defined for points in a neighborhood $U \subseteq \Sigma$ of x_*, where x_* is the intersection point of the limit cycle Γ with the surface Σ. Let x be such a point in a neighborhood. Then the Poincaré map $P: U \to \Sigma$ is defined by $P(x) = \phi_\tau(x)$, where τ is the time for the trajectory with initial point x to return to Σ for the first time. Note that τ can depend on the point x and may not be equal to the period T of the periodic orbit Γ. For the Poincaré map of the kicked harmonic oscillator of Section 1.2, $\tau = T$ for all points in the Poincaré section. This need not always be the case, but it is true that $\tau \to T$ as $x \to x_*$.

It is clearly true that x_* is a fixed point of the map P. We can characterize the fixed points and stability of maps in a manner similar to that of flows. In what follows we continue to denote a general mapping with the symbol P. This is not to imply that the results only apply to a Poincaré map, but rather to suggest to the reader this important application.

We have studied extensively the behavior of some one-dimensional maps already. In the multidimensional case the meaning of fixed points and periodic points is the same. We refer to the sequence of points $P(x), P^2(x), \ldots$ as the orbit of the point x under the map P. In the case that P denotes a Poincaré map the orbit of x consists of the intersections of the trajectory of x, $\phi_t(x)$, with the Poincaré section Σ.

A map P can be linearized about a fixed point x_* in a manner similar to that used for a flow. We examine the eigenvalues of $DP(x_*)$ and define the linear subspaces E_s, E_c, and E_u as the span of eigenvectors corresponding to eigenvalues with modulus less than 1, equal to 1, or greater than 1, respectively. There are corresponding manifolds for the nonlinear map P at x_*, tangent to the linear subspaces of $DP(x_*)$. When comparing maps with flows, it should be noted that for maps it is the modulus of the eigenvalues in comparison to 1 that decides stability, while for flows it is the real part of the eigenvalues in comparison to zero. For maps, an orbit is a sequence of points, whereas for flows a trajectory is a continuous curve.

As an example we consider the differential system

$$\begin{aligned} \dot{x} &= -y + \mu x - \mu x \sqrt{x^2 + y^2}, \\ \dot{y} &= x + \mu y - \mu y \sqrt{x^2 + y^2}. \end{aligned} \tag{5.18}$$

This system is more easily analyzed by using a polar coordinate system, $x = r\cos\theta$ and $y = r\sin\theta$. Then (5.18) becomes

$$\begin{aligned} \dot{r} &= \mu r(1 - r), \\ \dot{\theta} &= 1. \end{aligned} \tag{5.19}$$

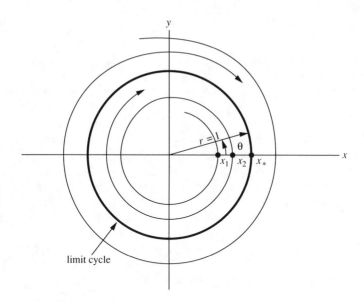

FIGURE 5.8   Trajectories for $\mu>0$ of system (5.18) or (5.19) near the limit cycle.

It is then elementary to find the solution

$$\phi_t(r_0, \theta_0) = \left(\frac{r_0}{r_0 - (r_0 - 1)e^{-\mu t}}, t + \theta_0\right). \tag{5.20}$$

The $r = 1$ circle in the plane is an attracting (repelling) trajectory or limit cycle of period 2π for $\mu > 0$ ($\mu < 0$). See Fig. 5.8.

As a Poincaré section we take the "plane" (line) Σ defined by points $y = 0$, $x > 0$, i.e., the $\theta = 0$ ray. The Poincaré map is one-dimensional and is given by

$$P(x) = x/(x - (x - 1)e^{-2\pi\mu}). \tag{5.21}$$

The map P of (5.21) clearly has a fixed point at $x = 1$ reflecting the intersection of the stable circular trajectory at $r = 1$. The linearization of P at 1 is

$$DP(1) = \frac{dP}{dx}\bigg|_{x=1} = e^{-2\pi\mu}. \tag{5.22}$$

Note that $DP(1) > 1$ when $\mu < 0$, implying instability. Similarly, $DP(1) < 1$ when $\mu > 0$, implying stability. In this case we can also linearize about the limit cycle solution by letting $\delta(t) = r(t) - 1$ and keeping only lowest order terms. We obtain

$$\dot{\delta} = -\mu\delta,$$
$$\dot{\theta} = 1, \tag{5.23}$$

giving the linearized flow

$$D\phi_t(\delta_0, \theta_0) = (\delta_0 e^{-\mu t}, t + \theta_0). \tag{5.24}$$

Evaluating at $t = 2\pi$ and $\theta = 0$ then gives $DP(1) = e^{-2\pi\mu}$ as in (5.22).

5.4 Center Manifolds

We have seen in previous chapters that systems can approach chaos through a series of bifurcations. In flows, as with maps, the nature of the bifurcations is determined by the local nature of the vector field $\mathbf{f}_\mu(\mathbf{x})$ near the bifurcation point in $(\mathbf{x}, \mu)$ space. To be even more specific, the nature of the bifurcation is determined by the dynamics on the center manifold at the bifurcation point. With the goal in mind of understanding these bifurcations we now attempt to isolate the dynamics on the center manifold.

As an example to illustrate the fundamental ideas we use the Lorenz system. This is our first encounter with this system in this book, but the Lorenz equations have been studied extensively and serve as one of the standard paradigms for chaotic dynamics. A good introduction to the Lorenz system is given in Guckenheimer and Holmes (1983, Section 2.3), and it is their discussion of center manifolds that we follow here. The Lorenz dynamical system is given by

$$\begin{aligned}
\dot{x} &= \sigma(y - x), \\
\dot{y} &= \rho x - y - xz, \\
\dot{z} &= -\beta z + xy,
\end{aligned} \qquad (5.25)$$

where $(x, y, z) \in \mathbb{R}^3$ and $\sigma, \rho, \beta > 0$. We are interested in the fixed point $(x, y, z) = (0, 0, 0)$. At the fixed point

$$D\mathbf{f} = \begin{pmatrix} -\sigma & \sigma & 0 \\ \rho & -1 & 0 \\ 0 & 0 & -\beta \end{pmatrix}. \qquad (5.26)$$

As eigenvalues of (5.26) we find

$$-\beta, \quad \lambda_\pm \equiv -(\sigma + 1)\left[\tfrac{1}{2} \pm \tfrac{1}{2}\sqrt{1 - 4\sigma(1 - \rho)/(\sigma + 1)^2}\right]. \qquad (5.27)$$

At $\rho = 1$ the eigenvalue $\lambda_- = 0$. So as ρ passes through 1, there is a bifurcation with a change of stability. As eigenvectors we take

$$\begin{pmatrix} 0 \\ 0 \\ 1 \end{pmatrix}, \quad \begin{pmatrix} -(1 + \lambda_+)/\rho \\ -1 \\ 0 \end{pmatrix}, \quad \begin{pmatrix} (1 + \lambda_-)/\rho \\ 1 \\ 0 \end{pmatrix}, \qquad (5.28)$$

corresponding, respectively, to the eigenvalues in (5.27). At $\rho = 1$ these eigenvectors become, respectively,

$$\begin{pmatrix} 0 \\ 0 \\ 1 \end{pmatrix}, \quad \begin{pmatrix} \sigma \\ -1 \\ 0 \end{pmatrix}, \quad \begin{pmatrix} 1 \\ 1 \\ 0 \end{pmatrix}, \qquad (5.29)$$

with the last vector in (5.29) tangent to the center manifold at the bifurcation point, since it corresponds with a zero eigenvalue.

We call the vectors in (5.28) $\mathbf{w}$, $\mathbf{v}$, and $\mathbf{u}$, respectively, and make a change of coordinates from the basis $\{\hat{\mathbf{x}}, \hat{\mathbf{y}}, \hat{\mathbf{z}}\}$ to $\{\mathbf{u}, \mathbf{v}, \mathbf{w}\}$. From (5.28) we have

$$
\begin{aligned}
\mathbf{w} &= \hat{\mathbf{z}}, \\
\mathbf{v} &= -(1+\lambda_+)\hat{\mathbf{x}}/\rho - \hat{\mathbf{y}}, \\
\mathbf{u} &= (1+\lambda_-)\hat{\mathbf{x}}/\rho + \hat{\mathbf{y}},
\end{aligned}
$$

and so the coordinates are given by

$$
\begin{pmatrix} x \\ y \\ z \end{pmatrix} = \begin{pmatrix} (1+\lambda_-)/\rho & -(1+\lambda_+)/\rho & 0 \\ 1 & -1 & 0 \\ 0 & 0 & 1 \end{pmatrix} \begin{pmatrix} u \\ v \\ w \end{pmatrix} \tag{5.30}
$$

with inverse

$$
\begin{pmatrix} u \\ v \\ w \end{pmatrix} = \begin{pmatrix} \rho/(\lambda_- - \lambda_+) & -(\lambda_+ + 1)/(\lambda_- - \lambda_+) & 0 \\ \rho/(\lambda_- - \lambda_+) & -(\lambda_- + 1)/(\lambda_- - \lambda_+) & 0 \\ 0 & 0 & 1 \end{pmatrix} \begin{pmatrix} x \\ y \\ z \end{pmatrix}. \tag{5.31}
$$

This gives as derivatives

$$
\begin{pmatrix} \dot{u} \\ \dot{v} \\ \dot{w} \end{pmatrix} = \begin{pmatrix} [(\lambda_+ + 1)(-xz - y + \rho x) - \rho\sigma(y - x)]/(\lambda_+ - \lambda_-) \\ [(\lambda_- + 1)(-xz - y + \rho x) - \rho\sigma(y - x)]/(\lambda_+ - \lambda_-) \\ xy - \beta z \end{pmatrix}. \tag{5.32}
$$

Substituting for (x, y, z) in terms of (u, v, w) from (5.30) gives

$$
\begin{aligned}
\dot{u} &= \frac{(\lambda_+ v - \lambda_- u)}{\lambda_+ - \lambda_-}\left[(\lambda_+ + 1)\left(\frac{w}{\rho} - 1\right) - \sigma\right] \\
&\quad + \frac{(v - u)}{\rho(\lambda_+ - \lambda_-)}\left[w(\lambda_+ + 1) - \sigma\rho(1 - \rho)\right], \\
\dot{v} &= \frac{(\lambda_+ v - \lambda_- u)}{\lambda_+ - \lambda_-}\left[(\lambda_- + 1)\left(\frac{w}{\rho} - 1\right) - \sigma\right] \\
&\quad + \frac{(v - u)}{\rho(\lambda_+ - \lambda_-)}\left[w(\lambda_- + 1) - \sigma\rho(1 - \rho)\right], \\
\dot{w} &= -\beta w + \frac{1}{\rho}[(\lambda_+ + 1)v^2 - (\lambda_+ + \lambda_- + 2)uv + (\lambda_- + 1)u^2].
\end{aligned} \tag{5.33}
$$

Equation (5.33) can be written symbolically in the form

$$
\begin{pmatrix} \dot{u} \\ \dot{v} \\ \dot{w} \end{pmatrix} = A \begin{pmatrix} u \\ v \\ w \end{pmatrix} + N(u, v, w), \tag{5.34}
$$

where A is a matrix and represents the linear part of the transformation, and $N(u, v, w)$ is a nonlinear vector. At the bifurcation point $\rho = 1$, (5.34) is

$$
\begin{pmatrix} \dot{u} \\ \dot{v} \\ \dot{w} \end{pmatrix} = \begin{pmatrix} 0 & 0 & 0 \\ 0 & -(\sigma + 1) & 0 \\ 0 & 0 & -\beta \end{pmatrix} \begin{pmatrix} u \\ v \\ w \end{pmatrix} + (u + \sigma v) \begin{pmatrix} -\sigma w/(\sigma + 1) \\ w/(\sigma + 1) \\ u - v \end{pmatrix}. \quad (5.35)
$$

Thus, we see that the center manifold is tangent to the u-axis. By setting $v = 0$, and $w = 0$, we see that the u-axis is not the invariant center manifold because of the u^2 term in $\dot{w}$, i.e., w does not stay equal to zero. We can push this noninvariance to a higher order by making a further coordinate transformation. We make a transformation to eliminate the quadratic term in $\dot{w}$. We let $\tilde{w} = w - au^2$ and determine the constant a to eliminate the u^2 term in $\dot{\tilde{w}}$. The new differential system then has the form

$$
\begin{pmatrix} \dot{u} \\ \dot{v} \\ \dot{\tilde{w}} \end{pmatrix} = A \begin{pmatrix} u \\ v \\ \tilde{w} \end{pmatrix} + \tilde{N}(u, v, \tilde{w}). \quad (5.36)
$$

The algebra involved in carrying out these calculations can become tedious but is easily handled on a computer with a symbolic manipulation program such as MACSYMA.

If we let $1 - \rho = \delta$, then on the u-axis ($v = 0$, $\tilde{w} = 0$) the dynamics is represented by the equation

$$
\dot{u} = \frac{\delta\sigma}{\sigma + 1} u - \frac{\sigma}{\beta(\sigma + 1)} u^3 + \text{higher order terms.} \quad (5.37)
$$

If we let $\tau = \sigma t/\beta(\sigma + 1)$ be a rescaled time and let $\mu = \beta\delta$, then (5.37) takes on the canonical form

$$
\frac{du}{d\tau} = \mu u - u^3 \quad (5.38)
$$

characteristic of pitchfork bifurcations for flows. We consider the canonical forms for bifurcations in a subsequent section.

We now wish to systemize this idea of reducing the dynamics near a bifurcation or equilibrium point to dynamics on a center manifold. Let us assume that the unstable manifold is empty and that our differential system (5.2) may be written in the form

$$
\begin{aligned}
\dot{\xi} &= A\xi + \mathbf{f}(\xi, \eta), \\
\dot{\eta} &= B\eta + \mathbf{g}(\xi, \eta),
\end{aligned} \quad (5.39)
$$

where ξ and $\mathbf{f}$ are n-vectors with A an $n \times n$ matrix having eigenvalues with real part equal to zero. Similarly, η and $\mathbf{g}$ are m-vectors and B is an $m \times m$ matrix having eigenvalues with negative real parts. The vectors $\mathbf{f}$ and $\mathbf{g}$ are assumed to vanish at the origin along with their Jacobian matrices $D\mathbf{f}$ and

$D\mathbf{g}$. Usually some transformations are necessary to put a differential system into this form and put the equilibrium point at the origin.

If it were now the case that $\mathbf{g}(\xi, 0) = 0$, then we would have the fortunate circumstance that the space of ξ is already a center manifold W_c specified by the equation $\eta = 0$. This is almost never the case, however. Instead W_c is specified by some "surface" over ξ, i.e., $\eta = \mathbf{h}(\xi)$ defines the center manifold W_c. The aim in what follows is to calculate the (vector) function $\mathbf{h}$.

To calculate the function $\mathbf{h}$ we project (5.39) onto the manifold W_c.

$$\dot{\xi} = A\xi + \mathbf{f}(\xi, \mathbf{h}(\xi)). \tag{5.40}$$

For the time derivative of η we have

$$\begin{aligned} \dot{\eta} &= D\mathbf{h}(\xi)\dot{\xi} = D\mathbf{h}(\xi)\big[A\xi + \mathbf{f}(\xi, \mathbf{h}(\xi))\big] \\ &= B\mathbf{h}(\xi) + \mathbf{g}(\xi, \mathbf{h}(\xi)). \end{aligned}$$

Thus we have the nonlinear partial differential equation to be satisfied by the function $\mathbf{h}(\xi)$:

$$D\mathbf{h}(\xi)\big[A\xi + \mathbf{f}(\xi, \mathbf{h}(\xi))\big] - B\mathbf{h}(\xi) - \mathbf{g}(\xi, \mathbf{h}(\xi)) = 0. \tag{5.41}$$

The structure of this equation is more explicit if we use index notation to denote derivatives and components. A comma denotes a partial derivative. Equation (5.41) in this notation becomes

$$h^i{}_{,j}(\xi)\big[A^j_k \xi^k + f^j(\xi, \mathbf{h}(\xi))\big] - B^i_j h^j(\xi) - g^i(\xi, \mathbf{h}(\xi)) = 0. \tag{5.42}$$

If the center manifold has more than one dimension, then (5.42) is a system of partial differential equations for the functions $h^i(\xi)$ that satisfy the boundary conditions $h^i(0) = 0$ and $h^i{}_{,j}(0) = 0$. Only in rare cases can this system be solved exactly. The solution strategy is to approximate the $h^i(\xi)$ by a Taylor series at the equilibrium point $\xi = 0$.

We saw in the example depicted in Fig. 5.3 that the center manifold need not be unique nor analytic. In such cases attempts to approximate $\mathbf{h}(\xi)$ by a Taylor expansion will be futile. As a representative example of such situations consider the system

$$\begin{aligned} \dot{x} &= -x^2, \\ \dot{y} &= -y + x^2. \end{aligned} \tag{5.43}$$

Clearly $(x, y) = (0, 0)$ is an equilibrium point and the x-axis is tangent to the center manifold at this point. It is also clear that the x-axis is not an invariant manifold because of the x^2 term in $\dot{y}$. However, any attempt to write $y = h(x)$ and then approximate $h(x)$ to increasingly high order by application of (5.42) will inevitably fail. This is because no y dependence exists in the equation for $\dot{x}$. Consequently, computation of $h(x)$ to any order makes no equation for $\dot{x}$. Consequently, computation of $h(x)$ to any order makes no improvement in the approximation of the equation of $\dot{x}$. In this case we can find a center manifold by solving for the curves in the (x, y) space directly. The differential system (5.43) becomes

$$\frac{dy}{dx} = \frac{y}{x^2} - 1, \tag{5.44}$$

which has the solution

$$y(x) = e^{-1/x}\left(\text{constant} - \int e^{1/x}\, dx\right). \tag{5.45}$$

With this cautionary glance into the possibilities for nonunique and non-analytic behavior on the center manifold behind us, let us consider a dynamical system for which (5.42) is the key to understanding the dynamics on the center manifold. Consider

$$\begin{aligned}
\dot{x} &= -y + xz, \\
\dot{y} &= x + yz, \\
\dot{z} &= -z - (x^2 + y^2) + z^2.
\end{aligned} \tag{5.46}$$

This system has a fixed point at $(x, y, z) = (0, 0, 0)$ and can immediately be written in the form (5.39).

$$\begin{pmatrix} \dot{x} \\ \dot{y} \end{pmatrix} = \begin{pmatrix} 0 & -1 \\ 1 & 0 \end{pmatrix} \begin{pmatrix} x \\ y \end{pmatrix} + \begin{pmatrix} xz \\ yz \end{pmatrix},$$
$$\dot{z} = (-1)z + (z^2 - x^2 - y^2). \tag{5.47}$$

At the equilibrium point the stable manifold is the z-axis, and the center manifold is tangent to the (x, y) plane. We identify the functions in (5.39) with the specific case of (5.46):

$$\xi = \begin{pmatrix} x \\ y \end{pmatrix}, \quad \eta = z, \quad \mathbf{f} = \begin{pmatrix} xz \\ yz \end{pmatrix}, \quad g = (z^2 - x^2 - y^2).$$

In this case the center manifold is given by an equation of the form $z = h(x, y)$. Equation (5.42) is now a single equation for the function $h(x, y)$.

$$xh_{,y} - yh_{,x} + h(xh_{,x} + yh_{,y}) + h - h^2 + x^2 + y^2 = 0. \tag{5.48}$$

We guess that $h(x, y)$ is of the form

$$h(x, y) = ax^2 + by^2 + cxy + Ax^3 + By^3 + Cx^2y + Dxy^2 + \mathcal{O}(|\xi|^4). \tag{5.49}$$

Note that taking $h(x, y)$ to be at least quadratic is required by the conditions that $h(0, 0) = 0$ and $(h_{,x}(0), h_{,y}(0)) = 0$. Now substitute (5.49) into (5.48)

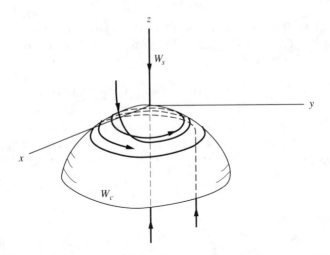

FIGURE 5.9   Sketch of the invariant manifolds for the system of (5.49).

and equate to zero the coefficients of each monomial up through third order. We find readily that

$$h(x,y) = -x^2 - y^2 + \mathcal{O}(|\xi|^4). \tag{5.50}$$

So that the center manifold is an inverted paraboloid, as shown in Fig. 5.9. Through third order the dynamics on the center manifold is determined by

$$\begin{pmatrix} \dot{x} \\ \dot{y} \end{pmatrix} = \begin{pmatrix} 0 & -1 \\ 1 & 0 \end{pmatrix} \begin{pmatrix} x \\ y \end{pmatrix} - \begin{pmatrix} x(x^2 + y^2) \\ y(x^2 + y^2) \end{pmatrix}. \tag{5.51}$$

The flow is most easily seen by converting to polar coordinates (r, θ) with $x = r \cos \theta$, $y = r \sin \theta$, which gives

$$\dot{r} = -r^3, \quad \dot{\theta} = 1$$

with solution

$$r = 1/\sqrt{2t + 1/r_0^2}, \quad \theta = t + \theta_0. \tag{5.52}$$

Thus the picture of the flow in the neighborhood of the origin is that the trajectories exponentially approach the center manifold in the z-direction and then slowly spiral in toward the origin. See Fig. 5.9.

It is a relatively simple matter to extend these computations to a dynamical system depending on parameters. Equation (5.39) becomes

$$\begin{aligned} \dot{\xi} &= A_\mu \xi + \mathbf{f}_\mu(\xi, \eta), \\ \dot{\eta} &= B_\mu \eta + \mathbf{g}_\mu(\xi, \eta), \\ \dot{\mu} &= 0, \end{aligned} \tag{5.53}$$

where $(\xi, \eta) \in \mathbb{R}^n \times \mathbb{R}^m$ and $\mu \in \mathbb{R}^k$. This system has a center manifold tangent to the (ξ, μ) space at $(\xi, \eta, \mu) = (0, 0, 0)$. We take $\eta = \mathbf{h}_\mu(\xi)$ and find, equivalent to (5.41), that

$$D_\xi \mathbf{h}_\mu(\xi)\left[A_\mu \xi + \mathbf{f}_\mu(\xi, \mathbf{h}_\mu(\xi))\right] - B_\mu \mathbf{h}_\mu(\xi) - \mathbf{g}_\mu(\xi, \mathbf{h}_\mu(\xi)) = 0. \qquad (5.54)$$

Again we seek the function $\mathbf{h}_\mu(\xi)$, but now as a power series in ξ and μ.

The Lorenz system as given in (5.33) is in the proper form to apply (5.54). We identify

$$\xi = u, \quad \eta = \begin{pmatrix} v \\ w \end{pmatrix}, \quad \mu = \delta = 1 - \rho.$$

To first order in δ

$$A_\delta = -\delta\sigma/(\sigma+1), \quad f_\delta = -\frac{\sigma w}{\sigma+1}\left[u\left(1 + \frac{2\sigma\delta}{(\sigma+1)^2}\right) + \sigma v\left(1 + \frac{\delta(\sigma^2 + 2\sigma - 1)}{(\sigma+1)^2}\right)\right],$$

$$B_\delta = \begin{pmatrix} -(\sigma+1) + \sigma\delta/(\sigma+1) & 0 \\ 0 & -\beta \end{pmatrix},$$

$$g_\delta = \begin{pmatrix} \frac{w}{\sigma+1}\left[u\left(1 - \frac{\delta(\sigma^2 - 2\sigma - 1)}{(\sigma+1)^2}\right) + v\sigma\left(1 + \frac{2\sigma\delta}{(\sigma+1)^2}\right)\right] \\ u^2\left(1 + \frac{\delta}{\sigma+1}\right) + uv(\sigma - 1)(1 + \delta) - v^2\sigma\left(1 + \frac{\sigma\delta}{\sigma+1}\right) \end{pmatrix}.$$

We now let

$$\mathbf{h}_\delta(u) = \begin{pmatrix} a_1\delta + b_1\delta u + c_1 u^2 \\ a_2\delta + b_2\delta u + c_2 u^2 \end{pmatrix}$$

and find by application of (5.54) that

$$a_1 = a_2 = b_1 = b_2 = c_1 = 0, \quad c_2 = 1/\beta.$$

This gives for the dynamics on the center manifold to lowest order

$$\dot{u} = -\frac{\delta\sigma}{\sigma+1}u - \frac{\sigma}{\sigma+1}(u^3/\beta),$$

and this result is consistent with (5.37).

We rescale as before and examine the bifurcation diagram of

$$\dot{u} = \mu u - u^3 = u(\mu - u^2). \qquad (5.55)$$

We have fixed points at $u = 0$, $u = \pm\sqrt{\mu}$ giving a bifurcation diagram as in Fig. 5.10 with the little arrows indicating stability type. This is an example of a pitchfork bifurcation for a flow. Bifurcations are discussed more completely in a following section.

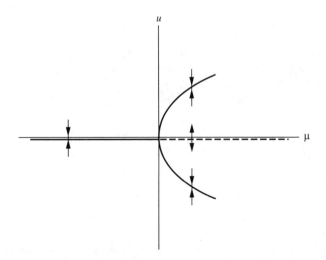

FIGURE 5.10   Pitchfork bifurcation diagram appropriate for the bifurcation of (5.55).

5.5 Normal Forms

Having studied the manner in which the interesting dynamics is reduced to
the center manifold, there remains yet one issue on the center manifold and
that is the choice of coordinates. We want the dynamics on this manifold to
be as "simple" as possible or to be in some sense "irreducible." This means
that in

$$\dot{\mathbf{x}} = \mathbf{f}_\mu(\mathbf{x}) \tag{5.56}$$

the vector field $\mathbf{f}_\mu(\mathbf{x})$ is as simple as possible. By simple we mean that $\mathbf{f}_\mu$ is
as close to being linear as is possible. Often symmetry considerations further
affect the form of the basis choice for expressing the vector field $\mathbf{f}_\mu(\mathbf{x})$. The
reduction to the simplest possible form is referred to as *reduction to normal
form*. This reduction proceeds only on the center manifold and, as noted
earlier in our discussion of center manifolds, is not always possible. In this
section, however, we confine our attention to polynomials, in which case the
reduction proceeds without difficulty. "Normal" is used in the sense of being
usual or common—the opposite of abnormal.

It is helpful in the following considerations to emphasize a more geometric
viewpoint of (5.56) than has heretofore been the case. In this light we think
of $\mathbf{f}_\mu(\mathbf{x})$ as the tangent vector to a trajectory in $\mathbb{R}^n$, where n is the dimension
of the center manifold. This tangent vector $\mathbf{f}_\mu(\mathbf{x})$ has a local coordinate
representation

$$\mathbf{f}_\mu(\mathbf{x}) = f_\mu^i(\mathbf{x})\frac{\partial}{\partial x^i} \equiv f_\mu^i(\mathbf{x})\partial_i, \tag{5.57}$$

where the components f_μ^i are functions of the coordinates x^i as noted, and
basis vectors are denoted as differential operators. We adopt from differential

geometry the viewpoint that vectors are operators. Acting on functions, for example, a vector gives the component of the function gradient along the direction of the vector. It is the polynomial coefficients of the vector operators, namely the $f_\mu^i(\mathbf{x})$, that we wish to be as close to being linear polynomials as possible.

As an example of the notation in (5.57), (5.55) would be written

$$\frac{du}{dt} = u(\mu - u^2)\partial_u, \tag{5.58}$$

and (5.51) would become

$$\frac{d\mathbf{x}}{dt} = [-y - x(x^2 + y^2)]\partial_x + [x - y(x^2 + y^2)]\partial_y. \tag{5.59}$$

A readable discussion of basis vectors as operators is found in Schutz (1980) with applications to dynamics in Rasband (1983) or Arnold (1978).

Once again, the issue we face in this section is that of choosing the coordinates in $\mathbb{R}^n$ such that the local representation of the vector field $\mathbf{f}_\mu(\mathbf{x})$ is simple. Note that the coefficients f^i are not only polynomials in the coordinates but can depend on map parameters such as the μ in (5.58). For reasons of brevity we suppress this dependence on parameters in what follows. However, we must remember this dependence since the major reason for this extensive analysis of the center manifold dynamics is to classify the bifurcations leading to chaos.

Let us assume that $\mathbf{f}(0) = 0$, i.e., $\mathbf{x} = 0$ is a fixed point. Furthermore, we assume that $\mathbf{f}(\mathbf{x})$ is smooth. As in most of what is done in nonlinear dynamics, we build on the linear situation. Denote

$$\mathbf{L}(\mathbf{x}) = D\mathbf{f}(0)[\mathbf{x}] \tag{5.60}$$

as the linear part of $\mathbf{f}(\mathbf{x})$. As an example, in (5.59) $\mathbf{L}(\mathbf{x}) = -y\partial_x + x\partial_y$.

Corresponding to $\mathbf{L}(\mathbf{x})$ is an adjoint map $\tilde{L}$ operating on vector fields. This map is induced by the Lie bracket with $\mathbf{L}(\mathbf{x})$. Let $\mathbf{v}$ be an arbitrary vector field, then

$$\tilde{L}[\mathbf{v}] = [\mathbf{v}, \mathbf{L}],$$

where $[\,,\,]$ denotes the usual Lie bracket. We recall that the Lie bracket of two vector fields is defined to be the vector field given by $[\mathbf{v}, \mathbf{u}] = (v^i u^j_{,i} - u^i v^j_{,i})\partial_j$, where commas denote partial differentiation. We shall see that the adjoint map is useful and natural in the context of our coordinate transformations.

Let $F^{(k)}$ denote the finite dimensional linear vector space over $\{\partial_i\}$ whose coefficients are homogeneous polynomials of degree k. That is to say, if $\mathbf{v} \in F^{(k)}$, then $v^i(\mathbf{x})$ is a homogeneous polynomial of degree k. Then

$$\tilde{L}: F^{(k)} \to F^{(k)}.$$

That this is so becomes transparent when we write out this map in component form:

$$(\tilde{L}[\mathbf{v}])^i = ([\mathbf{v}, \mathbf{L}])^i = v^j\, \partial_j L^i - L^j\, \partial_j v^i = v^j L^i_{,j} - L^j v^i_{,j}. \tag{5.61}$$

Since $\mathbf{L} \in F^{(1)}$, we see that the combinations of components and derivatives in (5.61) is so as to preserve the polynomial degree. It is not necessarily the case that $\tilde{L}(F^{(k)}) = F^{(k)}$, and we denote the complement of $\tilde{L}(F^{(k)})$ as $G^{(k)}$. Then $F^{(k)} = \tilde{L}(F^{(k)}) \oplus G^{(k)}$.

The basic result of this section is that the part of $\mathbf{f}(\mathbf{x})$ lying in $F^{(k)}$ can be eliminated by a coordinate transformation up to a component in the complement of $\tilde{L}(F^{(k)})$. We now proceed to show this result.

Suppose that $\mathbf{f}(\mathbf{x})$ can be written in the form

$$\mathbf{f}(\mathbf{x}) = \mathbf{f}^{(1)}(\mathbf{x}) + \mathbf{f}^{(2)}(\mathbf{x}) + \cdots, \tag{5.62}$$

where each $\mathbf{f}^{(i)}(\mathbf{x})$ belongs to the corresponding $F^{(i)}$, and furthermore, for $i < k$, $\mathbf{f}^{(i)}(\mathbf{x}) \in G^{(i)}$. In other words the $\mathbf{f}^{(i)}(\mathbf{x})$ for $i < k$ are already in normal form, i.e., as simple as possible. The term $\mathbf{f}^{(k)}(\mathbf{x})$ is the first one we can hope to reduce. Note that $\mathbf{f}^{(1)}(\mathbf{x})$ is just $\mathbf{L}(\mathbf{x})$. Now we make the coordinate transformation

$$\mathbf{x} = \mathbf{y} + \mathbf{P}(\mathbf{y}), \tag{5.63}$$

where $\mathbf{P}(\mathbf{y})$ has components that are each a homogeneous polynomial of degree k, the coefficients of which are to be determined. We have

$$\dot{\mathbf{x}} = \dot{\mathbf{y}} + D\mathbf{P}(\mathbf{y})\dot{\mathbf{y}} = (I + D\mathbf{P}(\mathbf{y}))\dot{\mathbf{y}} = \mathbf{f}(\mathbf{y} + \mathbf{P}(\mathbf{y}))$$
$$= \mathbf{L}(\mathbf{y}) + \mathbf{f}^{(2)}(\mathbf{y}) + \cdots + \mathbf{f}^{(k)}(\mathbf{y}) + D\mathbf{L}(\mathbf{y})[\mathbf{P}(\mathbf{y})] + \mathcal{O}(|y|^{k+1}). \tag{5.64}$$

The terms of degree less than k in (5.62) are left unchanged by this transformation. In substituting for $D\mathbf{P}(\mathbf{y})\dot{\mathbf{y}}$, we need keep only $D\mathbf{P}(\mathbf{y})\mathbf{f}^{(1)}(\mathbf{y})$ for the order indicated. Thus the new terms in (5.64) of order k are

$$\mathbf{f}^{(k)}(\mathbf{y}) + D\mathbf{L}(\mathbf{y})[\mathbf{P}(\mathbf{y})] - D\mathbf{P}(\mathbf{y})[\mathbf{f}^{(1)}(\mathbf{y})].$$

Since $\mathbf{f}^{(1)}(\mathbf{y}) = \mathbf{L}(\mathbf{y})$, this new term in component form is given by

$$(L^i_{,j} P^j - P^i_{,j} L^j)(\mathbf{y}) = ([\mathbf{P}, \mathbf{L}](\mathbf{y}))^i,$$

and so the new term in (5.64) is just

$$\mathbf{f}^{(k)}(\mathbf{y}) + \tilde{L}[\mathbf{P}(\mathbf{y})]. \tag{5.65}$$

We then choose the coefficients in $\mathbf{P}(\mathbf{y})$ so that (5.65) lies in $G^{(k)}$, thus increasing by 1 the number of terms in (5.62) lying in the complement of $\tilde{L}(F^{(k)})$.

It is clear that if $\tilde{L}(F^{(k)}) = F^{(k)}$, then we can eliminate completely from $\mathbf{f}(\mathbf{x})$ any vector field component with coefficients being homogeneous polynomials of degree k.

We now illustrate and explore our result by examining a few examples. First, consider (5.58). In this case $\mathbf{L}(u) = \mu u \partial_u$, and it is clear that $\tilde{L}[\mathbf{v}] = (1 - k)\mathbf{v}$ for all $\mathbf{v} \in F^{(k)}$. Consequently, $\tilde{L}(F^{(k)}) = F^{(k)}$, $k > 1$. Thus we can by this procedure linearize the polynomial coefficient in (5.58) to arbitrary order. Linearizing the polynomial coefficient to arbitrary order corresponds in the one-dimensional case to approximating the center manifold by coordinate transformations to increasingly high order.

Now consider Equation (5.59) with $\mathbf{L}(\mathbf{x}) = -y\partial_x + x\partial_y$ or $Df(0) = \begin{pmatrix} 0 & -1 \\ 1 & 0 \end{pmatrix}$ with eigenvalues $\pm i$. The normal forms are determined by the linear operator $\mathbf{L}$, i.e., the complementary spaces $G^{(k)}$ are determined by $\mathbf{L}$. Within these spaces, however, it is possible to pick different bases, and there may be symmetry considerations leading to the preference of one choice over another. In this example $\mathbf{L}(\mathbf{x})$ is invariant under rotations in the plane, thus we may choose bases in $G^{(k)}$ with the same property.

With $\mathbf{L}(\mathbf{x}) = -y\partial_x + x\partial_y$ Eq. (5.61) becomes

$$(\tilde{L}[\mathbf{v}])^i = \begin{pmatrix} -v^2 + yv^1_{,x} - xv^1_{,y} \\ v^1 + yv^2_{,x} - xv^2_{,y} \end{pmatrix}. \tag{5.66}$$

Because quadratic terms are already absent from (5.59), we need only look at the vector fields with coefficients consisting of homogeneous polynomials of third degree. From (5.66) we make a table of the action of $\tilde{L}$ on the possible basis vectors in $F^{(3)}$. The basis set in $F^{(3)}$ consists of $\{x^3\partial_x, x^2y\partial_x, xy^2\partial_x, y^3\partial_x, x^3\partial_y, x^2y\partial_y, xy^2\partial_y, y^3\partial_y\}$.

Table 5.1: Action of $\tilde{L}$ on Basis Vectors in $F^{(3)}$ from Eq. (5.66)			
	x^3	x^2y	xy^2	y^3
v^1	$3x^2y\partial_x + x^3\partial_y$	$(2xy^2 - x^3)\partial_x + x^2y\partial_y$	$(y^3 - 2x^2y)\partial_x + xy^2\partial_y$	$-3xy^2\partial_x + y^3\partial_y$
v^2	$-x^3\partial_x + 3x^2y\partial_y$	$-x^2y\partial_x + (2xy^2 - x^3)\partial_y$	$-xy^2\partial_x + (y^3 - 2x^2y)\partial_y$	$-y^3\partial_x - 3xy^2\partial_y$

From Table 5.1 we can write the linear operator $\tilde{L}$ on $F^{(3)}$ as a matrix. It has the block form

$$\tilde{L} = \begin{pmatrix} A & -I \\ I & A \end{pmatrix}, \tag{5.67}$$

where I is the 4×4 identity matrix and

$$A = \begin{pmatrix} 0 & -1 & 0 & 0 \\ 3 & 0 & -2 & 0 \\ 0 & 2 & 0 & -3 \\ 0 & 0 & 1 & 0 \end{pmatrix}. \tag{5.68}$$

If we assume $\mathbf{u}$ to be in the kernel of $\tilde{L}$, i.e., an eigenvector with zero eigen-value, and let $\mathbf{u} = (u_1, u_2)^t$, where the superscript t denotes transpose, we find the equation

$$(A^2 + I)u_i = 0, \qquad i = 1, 2. \tag{5.69}$$

As independent solutions of (5.69) in the kernel of $\tilde{L}$ and leading to the desired symmetry, we take

$$u_1 = \begin{pmatrix} 1 \\ 0 \\ 1 \\ 0 \end{pmatrix} \quad \text{and} \quad u_2 = \begin{pmatrix} 0 \\ 1 \\ 0 \\ 1 \end{pmatrix}. \tag{5.70}$$

Then as linearly independent vectors in the kernel of $\tilde{L}$, we take $(u_1, u_2)^t$ and $(-u_2, u_1)^t$.

These vectors then give as basis vectors for the complement $G^{(3)}$ of $\tilde{L}(F^{(3)})$

$$\mathbf{a}_1 = (x^2 + y^2)(x\partial_x + y\partial_y), \quad \mathbf{a}_2 = (x^2 + y^2)(x\partial_y - y\partial_x). \tag{5.71}$$

We see that (5.59) cannot be further simplified since it can be written in the form

$$\frac{d\mathbf{x}}{dt} = \mathbf{L} - \mathbf{a}_1.$$

Completely divorced from any connection with (5.59), we note that the complement $G^{(3)}$ spanned by the vectors $\mathbf{a}_1$, $\mathbf{a}_2$ of (5.71) depends only on the linear differential operator $\mathbf{L} = (x\partial_y - y\partial_x)$. This operator is invariant under rotations in $\mathbb{R}^2$, and the basis vectors of $G^{(3)}$ also have this symmetry property. This extension of the symmetry of the linear operator $\mathbf{L}$ to the complement space $G^{(k)}$ is true for general symmetries.

In preparation for consideration of bifurcations, it is time to put back the dependence on parameters. Thus note that a simple rescaling of the time would generalize $\mathbf{L}$ to the form $\mathbf{L} = \omega(x\partial_y - y\partial_x)$, giving eigenvalues $\pm i\omega$. To study bifurcations that would take place as the real part of the eigenvalues pass through zero, we should study an operator $\mathbf{L}$ with eigenvalues of the form $c\mu \pm i(\omega + \bar{c}\mu)$, where c and $\bar{c}$ are constants. Such an operator is

$$\mathbf{L} = (c\mu x - (\omega + \bar{c}\mu)y)\partial_x + ((\omega + \bar{c}\mu)x + c\mu y)\partial_y. \tag{5.72}$$

Exercise 5.4 shows that for $\mathbf{L}$ in (5.72), vector fields with quadratic polyno-mials can be completely eliminated. Thus to lowest order in μ, and from our previous example, we have that

$$\begin{aligned}
\frac{d\mathbf{x}}{dt} &= (c\mu x - (\omega + \bar{c}\mu)y)\partial_x + ((\omega + \bar{c}\mu)x + c\mu y)\partial_y \\
&\quad + a(x^2 + y^2)(x\partial_x + y\partial_y) + b(x^2 + y^2)(x\partial_y - y\partial_x) \\
&= \left[x(c\mu + a(x^2 + y^2)) - y(\omega + \bar{c}\mu + b(x^2 + y^2)) \right]\partial_x \\
&\quad + \left[x(\omega + \bar{c}\mu + b(x^2 + y^2)) + y(c\mu + a(x^2 + y^2)) \right]\partial_y
\end{aligned} \tag{5.73}$$

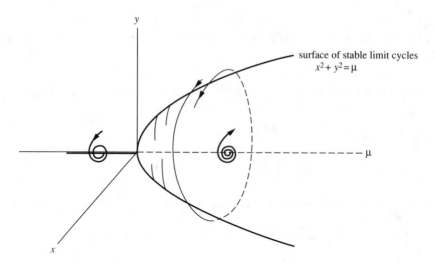

FIGURE 5.11   Bifurcation diagram for a Hopf bifurcation.

is the generic normal form for the dynamical equation having a two-dimensional center manifold at $\mu = 0$ with pure imaginary eigenvalues for **L**. The bifurcation that takes place at $\mu = 0$ is called a *Hopf bifurcation* and is discussed in the next section.

5.6 Bifurcations

As we have seen, bifurcations can be one of the important precursors to chaotic dynamics. Consequently, in this section we describe the more frequently occurring types of bifurcations. It is usually the case that the dynamics near the bifurcation is of interest, and thus a center manifold and normal form analysis is performed, if necessary. Thus, we do not study general theorems designed to tell us when bifurcations of a particular type can or will occur. For our purposes it is more important to recognize the possibilities and what the usual form of the vector field is like for the most common types of bifurcations. There are many excellent works discussing the issue of bifurcations in detail. For a general exposition, oriented toward differential equations, consult Chow and Hale (1982). For a computational viewpoint with applications see Kubíček and Marek (1983). In the context of group theory see Sattinger (1979), and for the Hopf bifurcation see Marsden and McCracken (1976).

We satisfy ourselves with a more descriptive account. Archetypical equations are given in Table 5.2 with bifurcation diagrams that are similar to the bifurcations for maps with similar names.

Comparing Tables 2.2 and 5.2 we see that the typical examples for maps and flows are very similar. The differences between the right-hand sides for the maps in Table 2.2 and differential equations in Table 5.2 are readily understood by the correspondence of the fixed point of a map with a vanishing for $\dot{x}$. Other differences in the bifurcation diagrams are only alternative ways for exchanges of stability to occur. Note that there is no bifurcation in one-dimensional flows analogous to the flip bifurcation for one-dimensional maps.

We now return to the single-parameter example of a Hopf bifurcation of the previous section. In Eq. (5.73) we set the constants $(\omega, a, b, c, \bar{c}) = (1, -1, 0, 1, 0)$ and obtain the archetypical differential system

$$\begin{aligned} \dot{x} &= -y + x(\mu - (x^2 + y^2)), \\ \dot{y} &= x + y(\mu - (x^2 + y^2)). \end{aligned} \qquad (5.74)$$

Table 5.2: Bifurcations for Flows in One Dimension

Name	Archetype Equation	Bifurcation Diagram
Saddle-node	$\dot{x} = \mu - x^2$	
Transcritical	$\dot{x} = \mu x - x^2$	
Pitchfork	$\dot{x} = \mu x - x^3$	

For the system of (5.74) the dynamics is best understood by changing to polar coordinates. One finds

$$\dot{r} = r(\mu - r^2), \qquad \dot{\theta} = 1.$$

For $\mu > 0$ there is a nontrivial periodic orbit at $r = \sqrt{\mu}$, where $\dot{r} = 0$. Linearization about $r = \sqrt{\mu}$ easily shows this circular orbit to be stable. The fixed point at $r = 0$ passes from being stable to unstable as μ increases through zero. As μ passes through zero, the new stable periodic orbit is born. Figure 5.11 is a sketch of the dynamics of (5.74).

If the sign on the cubic term in (5.74) is changed to $+$, or equivalently a is changed to $+1$ in (5.72), then we have a subcritical Hopf bifurcation as μ increases through zero. Exercise 5.5 supplies the details of this case.

5.7 Summary

In this chapter we have studied topics that form the foundation for our understanding of the transistion to chaos in differential systems. We first reviewed the concept of linearization and the dynamics induced by the linear part of the differential vector $\mathbf{f}_\mu(\mathbf{x})$ in (5.2). Building upon the discussion of invariant manifolds in the linear space, we considered some implications and applications of the center manifold theorem. We considered several examples illustrating the importance of the center manifold. On the center manifold, reduction to normal forms enables the identification of common routes of bifurcation. Bifurcations leading to chaos are considered for several examples in the following chapter.

Exercises

5.1 Make a qualitative sketch for the phase-space trajectories of the system

$$\ddot{x} + a^2 \sin x + \epsilon \dot{x}^2 = 0,$$

considering both positive and negative values of ϵ around zero. Identify the invariant manifolds of any fixed points. Check by actually solving for solution curves in phase space.

5.2 Obtain a qualitative sketch of the phase-space structure of the trajectories for the equation

$$\ddot{x} + \epsilon \dot{x} - x + x^3 = 0.$$

5.3 Analyze the dynamics on the center manifold for the system

$$\dot{u} = v,$$
$$\dot{v} = \beta u - u^2 - v.$$

(Hint:See Guckenheimer and Holmes (1983, Section 3.2).)

5.4 Show that the complementary space $G^{(2)}$ of $\mathbf{L}$ in (5.72) is empty.

5.5 Analyze the subcritical Hopf bifurcation at $\mu = 0$ for the differential system

$$\dot{x} = -y + x[\mu + (x^2 + y^2)],$$
$$\dot{y} = x + y[\mu + (x^2 + y^2)].$$

Make a sketch similar to Fig. 5.11 for this case.

CHAPTER SIX

NONLINEAR EXAMPLES WITH CHAOS

...to exaggerate the essential and purposely leave the obvious things vague
(Letters of Vincent Van Gogh)

In this chapter we consider some specific examples of differential systems in which chaotic dynamics is manifest for certain choices of the system parameters. There are a number of factors influencing our choice of examples. The first example we consider is a system of two coupled disk dynamos. For this system the physical model is rather straightforward and leads to behavior very much like the classical Lorenz system. Unfortunately, the art of dynamical modeling is often neglected in discussions of chaotic systems in spite of its crucial importance. For a recent book devoted to modeling for nonlinear systems see Beltrami (1987).

There are many systems in nature that are observed to be chaotic but for which no adequate physical model exists. Whether a model is adequate or not depends, of course, on the questions asked. The Lorenz system evolved out of an attempt to model the weather and is adequate to the extent that it is as unpredictable as the weather. The coupled disk dynamo system evolved from attempts to model the sudden reversals in the magnetic field of the earth. In the author's judgment the physical principles and basic equations for the disk dynamo are likely to be somewhat more familiar to the majority of the readers of this volume than the principles of fluid dynamics from which the Lorenz equations are derived. Thus, for pedagogical reasons, we consider the physical basis of the disk dynamo model in more detail than the Lorenz system, despite the fact that the chaotic behavior of the disk dynamo is less well understood. For a discussion of the physical basis of the Lorenz model we refer to the original discussion by Lorenz (1963) and for a more recent discussion to Bergé et al. (1984, Appendix D). As references on the disk dynamo we refer the reader to Cook and Roberts (1970) and Bullard (1978).

In this chapter we focus exclusively on three examples: the disk dynamo, the Lorenz system and the driven, damped pendulum. We conclude the chapter with a section on circle maps since such maps are related to the dynamics of the pendulum. Furthermore, the discussion of circle maps is a natural introduction into the following chapter on two-dimensional maps.

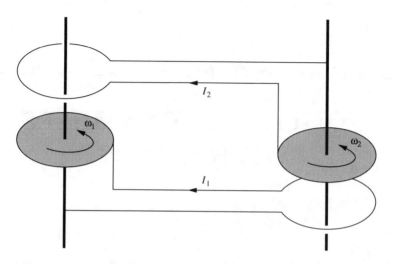

FIGURE 6.1   A coupled disk dynamo model.

6.1 The Disk Dynamo

Consider the simple system of two disk dynamos coupled as sketched in Fig. 6.1. Two disks of a highly conducting material are mounted on parallel shafts and to each is applied a constant torque Γ. Brushes are so positioned that current can flow radially in the disks and through the circular current loops placed above and below the disks as sketched in Fig. 6.1. There is an electric field in the frame of the conducting disk due to its motion in a magnetic field. This electric field drives a current in the attached current loop, thereby creating a magnetic field at the other disk. This coupling between disks provides the system with nonlinear coupling. Resistance provides dissipation and the applied torque supplies the drive.

For a perfect conductor the electric field in the frame of the moving conductor is given by

$$\mathbf{E} = -\frac{1}{c}\mathbf{v} \times \mathbf{B}. \tag{6.1}$$

We consider the directions indicated in Fig. 6.1 as positive and assume that at disk 1 a magnetic field is present of the form

$$\mathbf{B}_1 = B_z(r)\hat{z} + B_r(r)\hat{r}, \tag{6.2}$$

where r is a cylindrical coordinate measured from the rotation axis of the disk. The velocity of disk 1 is given by

$$\mathbf{v}_1 = \omega_1\hat{z} \times \mathbf{r} = \omega_1 r\hat{\phi}. \tag{6.3}$$

The emf generated between the axis and the edge by the motion of the disk in the magnetic field is given by

$$\mathcal{E}_1 = \frac{1}{c} \int_0^R (\mathbf{v}_1 \times \mathbf{B}_1) \cdot \mathbf{dr} = \frac{\omega_1}{c} \int_0^R r B_z(r) \, dr. \tag{6.4}$$

The magnetic field in (6.4) is due to the current in circuit 2, and so we may write

$$\mathcal{E}_1 = \omega_1 M I_2, \tag{6.5}$$

where M is a constant depending on the magnetic coupling of disk 1 with current loop 2.

We now consider the circuit equation for the potential drops in circuit 1 carrying current I_1.

$$\omega_1 M I_2 = I_1 R + L \dot{I}_1 + M_{12} \dot{I}_2, \tag{6.6}$$

where L is the self-inductance of the circuits and M_{12} is their mutual inductance. We obtain an analogous equation for the second disk and its circuit:

$$\omega_2 M I_1 = I_2 R + L \dot{I}_2 + M_{12} \dot{I}_1, \tag{6.7}$$

where we have assumed that the electrical parameters of the two circuits are identical.

To compute the mechanical properties of the rotating disks, note that for positive I_1 and I_2 the torque due to the magnetic field opposes the applied torque. For disk 1 it is given by

$$\frac{1}{c} I_1 \int_0^R r B_z(r) \, dr = I_1 I_2 M. \tag{6.8}$$

Assume that the moment of inertia of each disk about the axis of rotation is given by K. Then the mechanical equations are

$$\Gamma - M I_1 I_2 = K \dot{\omega}_1, \tag{6.9}$$
$$\Gamma - M I_1 I_2 = K \dot{\omega}_2 \tag{6.10}$$

Equations (6.6), (6.7), (6.9), and (6.10) form our dynamical system. It is, however, convenient to scale the variables in order to eliminate the appearance of all the physical constants. We let

$$I_i = x_i \sqrt{\Gamma/M}, \qquad \omega_i = y_i \sqrt{\Gamma L/KM}, \tag{6.11}$$

and rescale the time by defining

$$\tau = t \sqrt{\Gamma M/KL}. \tag{6.12}$$

With these changes in the independent and dependent variables the differential equations become

$$\dot{x}_1 + \nu\dot{x}_2 = -\mu x_1 + y_1 x_2,$$
$$\dot{x}_2 + \nu\dot{x}_1 = -\mu x_2 + y_2 x_1,$$
$$\dot{y}_1 = 1 - x_1 x_2,$$
$$\dot{y}_2 = 1 - x_1 x_2,$$

$$(6.13)$$

where $\mu = R\sqrt{K/LM\Gamma}$ and $\nu = M_{12}/L$. We note from the last two equations of (6.13) that the variable $\sigma \equiv y_1 - y_2$ is a constant.

In order to put (6.13) in standard form, we make a change of variables to

$$\bar{x}_1 = x_1 + x_2, \qquad \bar{x}_2 = x_1 - x_2, \qquad \bar{y} = (y_1 + y_2)/2. \qquad (6.14)$$

With this change the differential equations become

$$\dot{\bar{x}}_1(1+\nu) = -\mu\bar{x}_1 - \frac{\sigma}{2}\bar{x}_2 + \bar{y}\bar{x}_1,$$
$$\dot{\bar{x}}_2(1-\nu) = \frac{\sigma}{2}\bar{x}_1 - \mu\bar{x}_2 - \bar{y}\bar{x}_2,$$
$$\dot{\bar{y}} = 1 - \tfrac{1}{4}\bar{x}_1^2 + \tfrac{1}{4}\bar{x}_2^2.$$

$$(6.15)$$

We compute the fixed points in the usual way. It is convenient to define

$$\bar{x}_1^* = \eta + 1/\eta, \qquad \bar{x}_2^* = \eta - 1/\eta,$$

where $\sigma/\mu = \eta^2 - 1/\eta^2 = \bar{x}_1^*\bar{x}_2^*$. Then the two fixed points are given by

$$\mathbf{x}_\pm = \left(\pm\bar{x}_1^*, \pm\bar{x}_2^*, \frac{\mu}{2}(\eta^2 + 1/\eta^2)\right).$$

The Jacobian matrix at these fixed points is given by

$$Df(\mathbf{x}_\pm) = \begin{pmatrix} \frac{\mu\bar{x}_2^{*2}}{2(1+\nu)} & \frac{-\mu\bar{x}_1^*\bar{x}_2^*}{2(1+\nu)} & \frac{\pm\bar{x}_1^*}{(1+\nu)} \\ \frac{\mu\bar{x}_1^*\bar{x}_2^*}{2(1-\nu)} & \frac{-\mu\bar{x}_1^{*2}}{2(1-\nu)} & \frac{\mp\bar{x}_2^*}{(1-\nu)} \\ \frac{\mp\bar{x}_1^*}{2} & \frac{\pm\bar{x}_2^*}{2} & 0 \end{pmatrix}.$$

For $\nu = 0$ it is straightforward to find the three eigenvalues $\lambda = -2\mu, \pm i\omega$, where $\omega = \sqrt{\eta^2 + 1/\eta^2}$. Thus at $\nu = 0$ we have a two-dimensional center manifold. Following the procedure outlined in the previous chapter, we wish to find the corresponding eigenvectors so that we can make a coordinate change and achieve the splitting as assumed for (5.39). Since we have the situation of a stable manifold and a two-dimensional center manifold regardless of the values for μ and η, and because the algebra becomes very tedious, we make a convenient choice for μ and η so that the only parameter left is ν. Taking $\mu = 1$ and $\eta = 2$ we find

$$\bar{x}_1^* = \tfrac{5}{2}, \qquad \bar{x}_2^* = \tfrac{3}{2}, \qquad \text{and } \mathbf{x}_\pm = (\pm\tfrac{5}{2}, \pm\tfrac{3}{2}, \tfrac{17}{8}), \qquad (6.16)$$

and

$$Df(\mathbf{x}_\pm) = \begin{pmatrix} \frac{9}{8(1+\nu)} & \frac{-15}{8(1+\nu)} & \frac{\pm5}{2(1+\nu)} \\ \frac{15}{8(1-\nu)} & \frac{-25}{8(1-\nu)} & \frac{\mp3}{2(1-\nu)} \\ \frac{\mp5}{4} & \frac{\pm3}{4} & 0 \end{pmatrix}. \qquad (6.17)$$

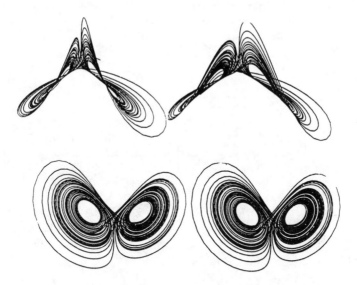

FIGURE 6.2   Stereo views of a trajectory of the coupled disk dynamo system.

The next thing to do is to move the origin of coordinates to one of the fixed points so that we may determine the structure of the center manifold in a neighborhood about the fixed point. We let

$$\xi_1 = \bar{x}_1 - \bar{x}_1^*, \quad \xi_2 = \bar{x}_2 - \bar{x}_2^*, \quad \zeta = \bar{y} - \frac{\mu}{2}\left(\eta^2 + \frac{1}{\eta^2}\right),$$

and substitute into to (6.15). We find

$$\begin{pmatrix} \dot{\xi}_1 \\ \dot{\xi}_2 \\ \dot{\zeta} \end{pmatrix} = Df \begin{pmatrix} \xi_1 \\ \xi_2 \\ \zeta \end{pmatrix} + \mathbf{N}(\xi_1, \xi_2, \zeta), \tag{6.18}$$

where Df is given in (6.17) with the upper sign choice and

$$\mathbf{N} = \begin{pmatrix} \xi_1 \zeta/(1+\nu) \\ -\xi_2 \zeta/(1-\nu) \\ -(\xi_1^2 + \xi_2^2)/4 \end{pmatrix}.$$

To first order in ν and with three significant figures in the numerical coefficients, the eigenvalues of Df are

$$\lambda_1 = -2 - 2.54\nu,$$
$$\lambda_{2,3} = -0.825\nu \pm i\left(\frac{\sqrt{17}}{2} - 1.31\nu\right). \tag{6.19}$$

At $\nu = 0$ there is a two-dimensional center manifold, and for $\nu < 0$ the fixed points have a two-dimensional unstable manifold. The structure of this system

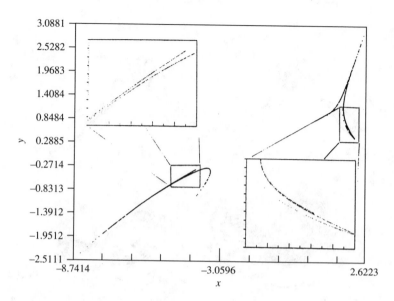

FIGURE 6.3   A Poincaré section plot for a trajectory similar to the one plotted in Fig. 6.2. The two pieces of the center manifold intersect the Poincaré section in the two pieces shown. Two parts of this section plane have been magnified in the inset to better show the interleaving on the center manifold.

indicates that at the fixed points a subcritical Hopf bifurcation takes place as the parameter ν passes through zero.

We pass now to a numerical integration of the system (6.15) with the parameters $\mu = 1$, $\eta = 2$, and $\nu = 0$. In Fig. 6.2 a trajectory segment for the time evolution is plotted in projection. The upper and lower figure pairs correspond to projection along different directions. The views are stereo projections and reveal the embedding of the center manifold in three-dimensional space. The reader can experience the stereo effect of these views by crossing his eyes so as to superimpose the images. Focusing on a pencil or similar object held a few centimeters off the page can sometimes be helpful. A necessary alternative for some readers will be a small stereo viewer.

Figure 6.3 shows a Poincaré section for similar parameters and initial data. The section plane for Fig. 6.3 is chosen to contain the line connecting the fixed points and to have a normal vector making and angle of 60° with the positive $\bar{y}$ axis. From these figures generated by numerical integration we gain some insight into the branched and folded nature of the center manifold surface.

The disk dynamo illustrates the flavor of the considerations that go into modeling a physical system. Usually numerous simplifying assumptions must be made in order to reduce the system to something tractable. But hopefully the system has not been simplified to the extent that some essential features of the physical system have been lost in the model. The extent to which any given model meets these two competing aims is usually not easy to determine. We have completely ignored any comparisons with data to assess the applicability of the disk dynamo to the system it was constructed to represent. We refer the reader to the previously cited literature for these issues.

Aside from issues of applicability to nature, the disk dynamo illustrates a Hopf bifurcation to an unstable limit cycle followed by chaos. The Poincaré plots indicate the complicated, interleaved structure of the center manifold that forms the attracting set for this system. Despite the bifurcation to the unstable limit cycle, the motion still remains bounded, indicating that global considerations can stabilize dynamics that is locally unstable. Global considerations form the subject of the next section.

6.2 Attracting Sets and Trapping Regions

From the eigenvalues of the Jacobian operator Df we obtain the essential information about the local stability of fixed points. However, local stability is not the whole story. A fixed point shown to be linearly unstable may be stabilized by nonlinear terms in a global sense. This is the case in the disk dynamo examined in the previous section. The bifurcation of the logistic map to a stable 2-cycle, just when the fixed point becomes unstable, is a simple example. Exponential growth can saturate and become bounded due to the nonlinearities in many ways. Other possibilities associated with intersections of stable and unstable manifolds were mentioned in Section 5.2.

An important tool for addressing questions of global stability consists of determining the attracting sets for a dynamical system or the existence of trapping regions. In Section 2.4 we considered attracting sets for maps. In analogous fashion a set A is called an *attracting set* for the dynamical system

$$\dot{\mathbf{x}} = \mathbf{f}(\mathbf{x}) \tag{6.20}$$

with flow designated by $\phi_t(\mathbf{x})$, if there is some neighborhood U of A such that $\phi_t(\mathbf{x}) \in U$ for $t \geq 0$ and $\phi_t(\mathbf{x}) \to A$ as $t \to \infty$ for all $\mathbf{x} \in U$. The *basin of attraction* of A is the union of all such neighborhoods U.

For some interesting dynamical systems it is possible to demonstrate the existence of a trapping region. A *trapping region* $D \subset \mathbb{R}^n$ is a closed, simply connected set, such that $\phi_t(D) \subset D$ for all $t \geq 0$. The existence of such a trapping region containing a fixed point ensures that all orbits in the neighborhood of a fixed point $\mathbf{x}_*$ are globally stable, regardless of the eigenvalues of $Df(\mathbf{x})$.

To demonstrate the existence of such a trapping region, it is sufficient to find a closed, connected set D on which the vector field $\mathbf{f}$ of (6.20) is always directed inward. The attracting set A for such a trapping region is given by

$$A = \cap \phi_t(D), \quad t > 0.$$

To find a trapping region is not always possible, but often an "energy-like" surface can be shown to be a trapping region. Consider a flow $\phi_t(\mathbf{x}) \subset \mathbb{R}^n$

determined by a system of differential equations of the form (6.20). Following an argument due to Lorenz (1963), we more specifically let x_i denote the ith component of $\mathbf{x}$, and consider a differential system of the form

$$\dot{x}_i = -\sum_j a_{ij}x_j + \sum_{j,k} b_{ijk}x_jx_k + c_i, \qquad (6.21)$$

where

$$\sum_{i,j} a_{ij}x_ix_j > 0 \quad \text{and} \quad \sum_{i,j,k} b_{ijk}x_ix_jx_k \equiv 0 \quad \text{for all} \quad \mathbf{x} \in I\!\!R^n. \qquad (6.22)$$

The matrices a_{ij}, b_{ijk}, and c_i are all constant. Consider a surface in $I\!\!R^n$ defined by

$$Q = \frac{1}{2}\sum_i x_ix_i. \qquad (6.23)$$

Then we let the points on the surface Q=constant evolve according to the flow $\phi_t(\mathbf{x})$ defined by (6.21) and examine the change in Q.

$$\frac{dQ}{dt} = \sum_i x_i\dot{x}_i = -\sum_{i,j} a_{ij}x_ix_j + \sum_i c_ix_i. \qquad (6.24)$$

Since a_{ij} is a positive definite matrix, as implied by the first of (6.22), a coordinate transformation can be performed such that a_{ij} is diagonal with its eigenvalues (all positive) along the diagonal. Assuming this diagonalization has taken place, (6.24) can be written in the form

$$\begin{aligned}\frac{dQ}{dt} &= -(\lambda_1x_1^2 + \lambda_2x_2^2 + \cdots + \lambda_nx_n^2) + c_1x_1 + \cdots + c_nx_n \\ &= -\sum_i\left[\left(\sqrt{\lambda_i}x_i - \frac{c_i}{2\sqrt{\lambda_i}}\right)^2 - \frac{1}{4}\frac{c_i^2}{\lambda_i}\right].\end{aligned} \qquad (6.25)$$

Equation (6.25) implies that if Q is taken large enough so that the "spherical" surface defined by Q through (6.23) encloses the ellipsoid with semiaxes c_i/λ_i, then the flow is directed inward everywhere on the surface defined by Q, and the volume enclosed by Q is a trapping region. The system is globally stable if the initial point for the flow is in the trapping region.

A related issue is associated with the contraction of volume by the flow. Consider a volume ΔV in the phase space of the dynamical system (6.20).

$$\Delta V = \Delta x_1\,\Delta x_2\cdots\Delta x_n.$$

As a function of time this volume will change as each point in it evolves according to (6.20):

$$\frac{d\,\Delta V}{dt} = \frac{d\,\Delta x_1}{dt}\Delta x_2\cdots\Delta x_n + \cdots + \Delta x_1\cdots\frac{d\,\Delta x_n}{dt}.$$

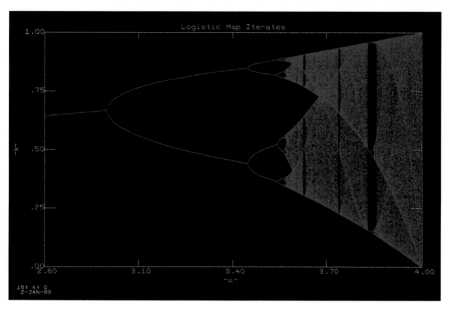

PLATE I   Iterates of the logistic map as a function of the map parameter μ for $2.8 < \mu < 4.0$.

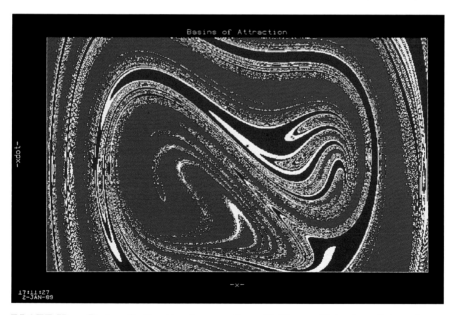

PLATE II   Basins of attraction for a nonlinear Duffing oscillator in a Poincaré section. The red and blue regions are for the fixed points that are marked with + signs. The white is the basin of attraction for a 6-cycle with section points marked by *.

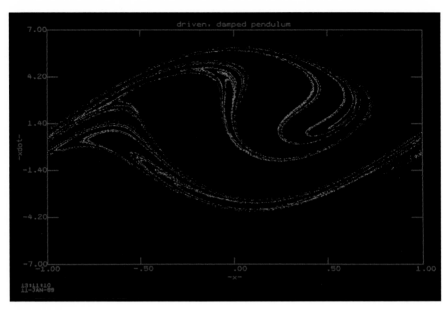

PLATE III   Poincaré section for a driven, damped pendulum in a running chaotic state.

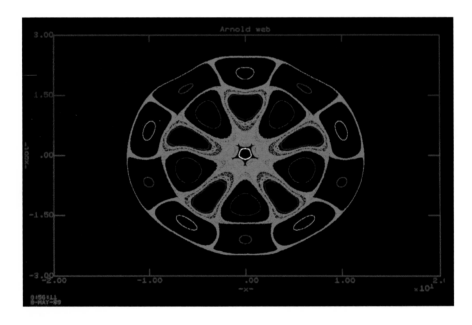

PLATE IV   The Arnold web for a perturbed Hamiltonian system.

As representative for Δx_i, the infinitesimal time derivative of Δx_1 is given by

$$\frac{d\,\Delta x_1}{dt} = \frac{\partial f_1}{\partial x_1}\Delta x_1,$$

and similarly for the other coordinate directions. Consequently,

$$\frac{d\,\Delta V}{dt} = (\nabla \cdot \mathbf{f})\,\Delta V. \tag{6.26}$$

From (6.26) we learn that if $\mathbf{f}$ in (6.20) has negative divergence at a point $\mathbf{x} \in I\!\!R^n$, then volumes are contracting at that point.

Frequently, systems are encountered for which $\nabla \cdot \mathbf{f} < 0$ everywhere, implying that under the flow phase-space volumes contract to zero. This means that any attracting set of such a system must have zero phase-space volume. In the next section we study the Lorenz system that contracts volume everywhere. The attracting set is a strange attractor with dimension $\simeq 2.02 <$ 3 having zero volume.

6.3 The Lorenz System

For an example of a system with a trapping region we consider once again the Lorenz system as given in (5.25). Comparing (5.25) with the general form (6.21), we identify

$$a_{ij} = \begin{pmatrix} \sigma & -\sigma & 0 \\ -\rho & 1 & 0 \\ 0 & 0 & \beta \end{pmatrix}, \tag{6.27}$$

with $b_{213} = b_{231} = -\frac{1}{2}$, $b_{312} = b_{321} = \frac{1}{2}$, and all other components of b_{ijk} along with c_i equal zero. However, a_{ij} of (6.27) is not positive definite. For this case this defect is easy to remedy by transforming to new variables and a modified system where all the conditions of (6.22) are satisfied. Let $z = \bar{z} + \rho + \sigma$. Then the new system of equations satisfies all conditions of (6.22) with a new a_{ij} given by

$$a_{ij} = \begin{pmatrix} \sigma & -\sigma & 0 \\ \sigma & 1 & 0 \\ 0 & 0 & \beta \end{pmatrix}$$

with the constant vector $c_i = (0, 0, -\beta(\rho + \sigma))^t$, where the superscript t denotes transpose. Consequently, for a large enough value of Q, the volume enclosed by the surface defined through (6.23) serves as a trapping region.

Consider the volume contraction of the Lorenz system. In this case we find that $\nabla \cdot \mathbf{f} = -(\sigma + 1 + \beta)$ independent of the point considered. Since σ and β are both positive, the phase-space volume is everywhere contracted in the Lorenz system. We have the interesting situation that a trapping region exists, and everywhere in this trapping region the flow contracts volumes. The attracting set for the Lorenz flow then must have zero volume in $I\!\!R^3$. If the attracting set is neither a fixed point nor a limit cycle, it must be a strange attractor since it occupies zero volume.

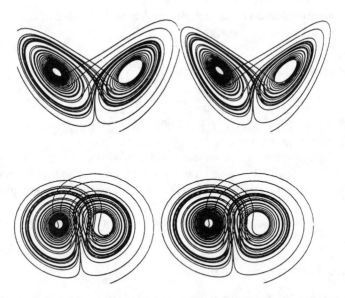

FIGURE 6.4   Stereo views of a trajectory segment on the center manifold for the Lorenz system. Both views are of the same trajectory but projected into different viewing planes.

The Lorenz system has been studied in considerable detail, and the behavior of the attracting sets as a function of the parameters is reasonably well understood. It would take us too far afield to go into all details of the analysis of the Lorenz system, but we recommend the monograph by Sparrow (1982) for a thorough discussion and for an extensive list of references.

Much of the understanding of the Lorenz attractor comes from numerical experiments, and consequently we turn to some numerical results for the Lorenz system of (5.25). We choose the parameter values $\sigma = 10$, $\beta = \frac{8}{3}$, and $\rho = 28$, which are the standard choices as first described by Lorenz (1963). Figure 6.4 shows a sample trajectory in stereo views.

Figure 6.5 shows a Poincaré section for a similar trajectory. One can see from the trajectory in Fig. 6.4 that the sheaves of the attracting set must be interleaved in a complicated way. However, the volume contraction is so strong for the Lorenz system that the Poincaré section of Fig. 6.5 reveals none of the complicated structure of the interleaving of surfaces forming the attractor, as is the case for the disk dynamo shown in the Poincaré section of Fig. 6.3. For our choice of parameters $d\,\Delta V/dt = -(\frac{41}{3})\,\Delta V$, and thus for unit time the initial volume is contracted by a factor $e^{(-41/3)} \simeq 10^{-6}$. Nevertheless, the chaotic nature of the Poincaré map can be easily demonstrated in a number of ways.

Following Lorenz (1963) we consider the surface on which $\dot{z} = 0$, given by $xy - \beta z = 0$. The maximum of z for every trajectory lies on this surface and we consider $z_{\max}$ at the nth intersection. Figure 6.6 shows a numerically generated plot of this function including the line with slope 1. Even on the

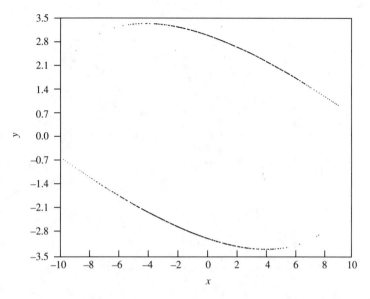

FIGURE 6.5   Poincaré section for the Lorenz attractor.

most cursory examination, one cannot help but notice the similarity of this
figure with the graph of the tent map in Fig. 2.1. The slope of the map in
Fig. 6.6 is everywhere greater than 1, and all of the arguments regarding the
chaotic nature of the tent map are applicable for the map of Fig. 6.6.

An alternative one-dimensional map that can be used to illustrate the

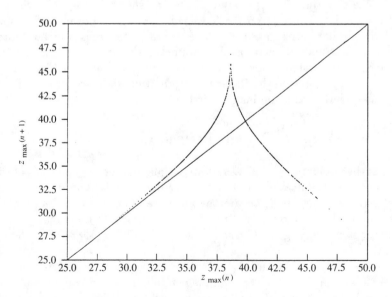

FIGURE 6.6   One-dimensional map considered by Lorenz.

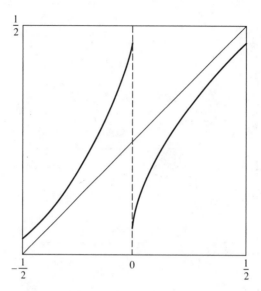

FIGURE 6.7   One-dimensional map constructed from the Poincaré map in the plane $z = \rho - 1$.

chaotic nature of the Lorenz attractor is obtained by projecting the points in one part of the Poincaré section in Fig. 6.5 back onto the line connecting the fixed points and then considering the map of points on this line, i.e., a point on this line at the $(n + 1)$th intersection as a function of the nth intersection point. The resulting map is similar to the one shown in Fig. 6.7, where the length of the interval between the fixed points has been normalized to unity.

Maps similar to that in Fig. 6.7 have been used to discuss the development of the Lorenz strange attractor as the parameter ρ is varied. For this discussion we refer the interested reader to Sparrow (1982, Chapter 3). The map in Fig. 6.7 is very similar to the Baker transformation discussed in Exercise 2.10 and consequently has a similar chaotic behavior.

6.4 The Damped, Driven Pendulum

Some of the standard dynamical systems displaying the phenomenon of chaos fall under the classification of nonlinear oscillators. The Van der Pol oscillator and the Duffing oscillator are the most well-known examples, and their dynamics has been thoroughly studied. As a reference to the Van der Pol and Duffing systems we suggest Guckenheimer and Holmes (1983) and references given therein.

In this section we focus our attention once again on a variation of the simple pendulum. We consider a simple pendulum that is damped and driven by an oscillatory torque. There has been considerable interest in this system

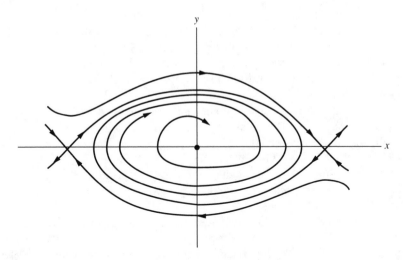

FIGURE 6.8   Phase curves of a damped simple pendulum.

because the differential equations also model the time dependence of the quantum phase difference of a Josephson junction. The differential equation of the damped, driven pendulum has also been modeled with an electronic circuit, which then serves as an analog device to explore extensively the dynamical behavior of this system. We refer specifically to D'Humieres et al. (1982), Huberman et al. (1980), and Pederson and Davidson (1981).

The differential equation for the driven, damped pendulum is given by

$$\ddot{x} + \Omega^2 \sin x = -\alpha \dot{x} + f \cos \omega t, \tag{6.28}$$

where Ω represents the natural frequency of the pendulum for small oscillations, α measures the strength of the damping, and f measures the strength of the drive.

With α and f both zero the phase curves for the simple pendulum are familiar and given in Fig. 5.5, where the x-axis has been extended beyond $\pm\pi$ in the usual way. We reduce (6.28) to standard, nonautonomous form by letting $y = \dot{x}$. This gives

$$\dot{x} = y,$$
$$\dot{y} = -\Omega^2 \sin x - \alpha y + f \cos \omega t. \tag{6.29}$$

If $f = 0$, the fixed points remain the same as for the simple pendulum, but now the phase curves have the qualitative form given in Fig. 6.8. Note that the damping forces almost all orbits to approach the origin.

If we add the driving term in (6.29), then it is clear that the fixed points of Figs. 5.5 and 6.8 no longer remain fixed. However, if the amplitude f of the driving term is small, then we expect on physical grounds a limit cycle around the origin $(x, y) = 0$ with a frequency equal to the driving frequency ω. On the limit cycle drive and damping exactly balance. If the amplitude of the drive is increased, then the dynamics of the system will begin to feel the influence of the former unstable fixed points. For a sufficiently large amplitude for the drive, we can expect to see interesting effects due to the nonlinear nature of the system.

The numerical investigation of such systems proceeds as with the previous systems studied in this chapter, except that making a Poincaré plot is especially simple. For a forced oscillator we can choose the $(x, \dot{x})$ phase plane as the Poincaré section. The return time to the Poincaré section is then no longer indeterminate, but rather just equal to the period of the forcing term T. In other words the points in the Poincaré section plane are the $(x, \dot{x})$ points at $t = 0, T, 2T, \ldots$.

The usual way to explore the dynamical behavior of a system such as (6.29) is to fix α and f and then vary the driving frequency. A larger response of the oscillator to the driving term is to be expected when the driving frequency is near to the natural frequency Ω of the simple pendulum.

In any system like the driven, damped pendulum, the various choices of the parameters can be very important for determining the behavior of the system. If the construction or operation of a device is at issue, then regions in parameter space that lead to chaotic behavior are generally to be avoided. If, on the other hand, some naturally occurring chaotic system is being modeled, it may be exactly the chaotic regions in parameter space that are of primary interest. It is safe to say that even for the most thoroughly studied systems, a complete map of parameter space with a catalog of system behavior is not available. The driven, damped pendulum is no exception. We report here some of the features that are known about this system in order to illustrate the questions being asked and also try to indicate those that remain open. The results we report are mostly taken from the research articles referred to at the beginning of this section. We hasten to add that not all discrepancies have been resolved between the numerical investigations of Pederson and Davidson (1981) and the experimental investigations reported in Huberman et al. (1980) and D'Humieres et al. (1982).

The most prevalent orbits are mode-locked orbits where the pendulum oscillates at the frequency of the driving force. These orbits are stable attractors and in the phase space of Figs. 5.5 and 6.8 are period-1 limit cycles. However, for certain choices of the parameters other attracting sets for the system of (6.29) exist. Limit cycles with periods other than one are readily found. In addition, chaotic attractors abound. Figure 6.9 is an attempt to delineate regions in parameter space leading to particular kinds of dynamical behavior. Similar "bifurcation diagrams" are published in the other references cited for this section.

As complicated as such diagrams appear, suggesting a richness of behavior, they are infact grossly incomplete. Figure 6.9 was constructed by numerically integrating (6.29), and, of course, numerical integrations are al-

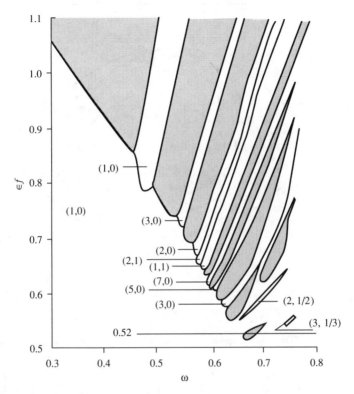

FIGURE 6.9   Shows dynamical behavior for a region of parameter space. For this plot $\alpha = 0.2$ in (6.29). The cross-hatched region represents chaos; the hatched region is for complicated periodic behavior and the indexing is for mode-locked states. From Pederson and Davidson (1981).

ways limited by finite accuracy. Furthermore, there is the obvious restriction that Fig. 6.9 attempts to be nothing more than a slice through the parameter space of (ω, f, α) at the fixed value $\alpha = 0.2$. However, diagrams such as Fig. 6.9 fail to represent the fact that there may be many attracting sets, each with its own basin of attraction. Pederson and Davidson (1981) were careful to state that they always chose the initial state to be at the origin in phase space. A different choice of initial conditions can give a completely different picture for the attracting sets at the same choices for the parameters (ω, f, α).

As examples illustrating the types of solutions obtained and the points mentioned previously, we offer the examples depicted in Figs. 6.10–6.12. Note that the actual angular position of the pendulum is the x-value in the figures, mod(an odd multiple of π); physically, the angular position of the pendulum always falls in the interval $(-\pi, \pi)$. We always choose the natural frequency $\Omega = 1$ and, as for the cases of Fig. 6.9, we choose $\alpha = 0.2$. With $f = 0.52$ we scan across the frequency ω along the horizontal line drawn on Fig. 6.9. Most values of ω simply lead to period-1 limit cycles. Figure 6.10 shows, however, a

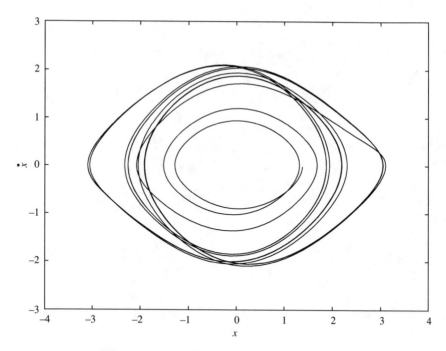

FIGURE 6.10   Trajectory segment for system (6.29) with $\alpha = 0.2$, $f = 0.52$, $\omega = 0.689$, and $x(0) = 0.8$, $\dot{x}(0) = 0.8$.

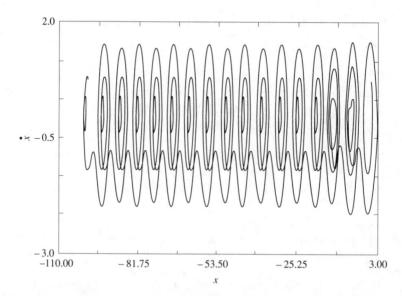

FIGURE 6.11   Trajectory segment for a period-3 limit cycle. Parameters and initial conditions are the same as in Fig. 6.10 except $\omega = 0.694$.

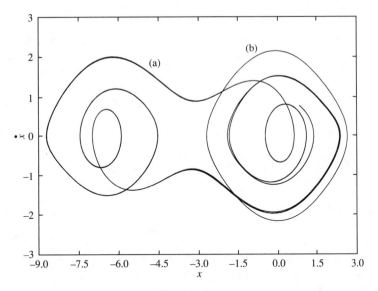

FIGURE 6.12   (a) Trajectory segment for a period-5 limit cycle with $\omega = 0.668$ and other parameters and initial conditions as in Fig. 6.10. (b) A period-1 limit cycle with same parameters but with the initial conditions $x(0) = -0.8$, $\dot{x}(0) = 0.1234$.

period-3 limit cycle for the frequency and initial conditions as indicated in the figure caption. A period-3 limit cycle, but now a so-called running solution is depicted in Fig. 6.11.

The reader should note that none of these attracting sets are noted in Fig. 6.9; the only thing being different is the initial point in phase space. Figure 6.12 reinforces this point concerning the role played by the basin of attraction wherein a period-5 limit cycle and a period-1 limit cycle are obtained for exactly the same choice of system parameters with only the initial conditions differing. This case amply illustrates that there can be many attracting sets, each with its own basin of attraction. The ultimate state of such an oscillator depends not only on the values of the system parameters but also on its initial condition.

The dependence of the final state on initial conditions is illustrated with particular force in Plate II. The colors in this plate show the basins of attraction for a nonlinear Duffing oscillator

$$\ddot{x} - \frac{1}{2}x(1 - x^2) = -\alpha\dot{x} + f\cos\omega t, \qquad (6.30)$$

which is similar to (6.28). For the purpose of illustrating the complex nature of basins of attraction, this system has the advantage that it is easy to find parameter values where there exist a small number (greater than one) of simple attracting sets in a Poincaré section. For Plate II $\alpha = 0.15$, $f = 0.1$, and $\omega = 0.833$. The attracting sets are two fixed points and a 6-cycle. All initial points in the blue area of Plate II are attracted to a fixed point in the blue area. Likewise, all initial points in the red area are attracted to the fixed point marked with a plus sign in the red area. All initial points in the white area are attracted to the 6-cycle whose section points are marked with asterisks in Plate II. Moon and Li (1985) reported on this system and showed the basin boundaries to be fractal.

Figure 6.9 shows many regions of chaos. All forms of chaos are not the same, nor is the manner in which the system becomes chaotic always the same; the variety is bewildering. It is interesting to note that with the parameter choices $\Omega = 1$, $\alpha = 0.25$, $f = 0.68$ and with the initial condition $x(0) = -1.5$, $\dot{x}(0) = 1.5$, there is period-2 limit cycle at $\omega = 0.674$. With a careful decrease in the value of ω, a period-doubling transistion to chaos occurs.

Finally, Plate III shows a Poincaré section for a completely chaotic state with $\Omega = 2.53$, $\alpha = 0.25$, $f = 3.8$, and $\omega = 1.62$. This is an example of a running chaotic state where the pendulum is frequently going over the top. Notice the self-similar structure of this attractor as each fold curls in on the one above it. Each fold contains a detailed inner structure, the resolution being limited by the size of the computational window examined and the number of intersection points. The regions where there are a larger number of points indicate regions of higher probability for finding the system at any time nT, where n is some positive integer.

The numerical experiments that we have described in this section for the driven, damped pendulum illustrate some of the complexities contained in such nonlinear oscillators. The dynamic behavior and attracting sets varies greatly with differing choices for the system parameters, and further with differing choices for initial conditions. Much remains to be done in understanding the behavior of such systems; presently, we are at the stage of mostly cataloging.

6.5 The Circle Map and the Devil's Staircase

The driven, damped pendulum exhibits an enthralling richness of behavior as a function of the mapping parameters in (6.29). In Huberman et al. (1980) the authors report the observation of period-doubling transistions to chaos, but this is clearly not the only route that this system may follow to chaos. In this section we describe a further route to chaos through quasiperiodicity and mode locking. This route to chaos also exhibits universal behavior. A readable, nontechnical article discussing these topics is Bak (1986). For research literature we refer the interested reader to Jensen et al. (1983, 1984), Bohr et al. (1984), Cvitanović et al. (1985), and Feigenbaum et al. (1982).

The key point in the dynamics of the damped, driven pendulum is that the strong damping causes the Poincaré map associated with (6.29) to reduce to a one-dimensional map on the circle. It is convenient and customary to write these so-called circle maps in the form

$$\theta_{n+1} = f(\theta) = \theta_n + \Omega + g(\theta_n), \qquad (6.31)$$

where

$$g(\theta_n) = g(\theta_n + 1). \tag{6.32}$$

Note that the angle θ has been normalized so that its range is $[0,1)$ instead of $[0, 2\pi)$. The function $g(\theta)$ represents the nonlinear part of the mapping, and Ω gives the frequency of the map in the absence of the nonlinear term.

In the driven, damped pendulum there are two frequencies in the system: the natural frequency of the pendulum and the driving frequency. When the coupling is weak, i.e., when f in (6.29) is small, the motion is quasiperiodic. As the strength of the coupling increases, i.e., as f increases, then the phenomenon of *mode locking* can occur. In (6.31) mode locking manifests itself through periodicity. That is after q iterations of the map (6.31), the new angle differs from the initial value by an integer p.

$$\theta_{n+q} = \theta_n + p. \tag{6.33}$$

The *winding number* of the map is a convenient way of characterizing the behavior of (6.31) under iteration. The winding number is defined by

$$W = \lim_{n \to \infty} [(f^n(\theta) - \theta)/n]. \tag{6.34}$$

The quantity W represents the average increase in the angle θ per unit time (average frequency) and for a phase-locked map as in (6.33), $W = p/q$. Thus a mode-locked state is when W equals a rational number. We refer to quasiperiodic states as those with irrational winding numbers. A chaotic state is when the winding number is not defined.

The map that serves as the archetype for understanding circle maps such as (6.31) is the map obtained when $g(\theta)$ is a simple sine function:

$$\theta_{n+1} = f(\theta_n) = \theta_n + \Omega - \left(\frac{k}{2\pi}\right) \sin 2\pi\theta_n. \tag{6.35}$$

In this map (6.35) the quantity k is the parameter that measures the strength of the nonlinearity. For $k = 0$ it is clear that $W = \Omega$. Thus when $k = 0$ we formally have mode-locking whenever Ω in (6.35) equals a rational number. The rationals are, however, a set of measure zero in the interval $[0,1]$, and so almost all maps are quasiperiodic for $k = 0$.

As k increases, however, the measure of the set of values Ω for which W is a rational number increases until at $k = 1$ the measure of the values of Ω for which the motion is quasiperiodic becomes zero. This happens because the set of values of Ω for which W is irrational has become a Cantor set.

In order to give an indication of the manner in which the width of a given resonance interval, $\Delta\Omega(p/q)$, is calculated, we look at $\Delta\Omega(0/1)$. This resonance corresponds to a fixed point in the map. We denote the end of the interval Ω_* and the fixed point θ at this value of Ω as θ_*. Then the equations

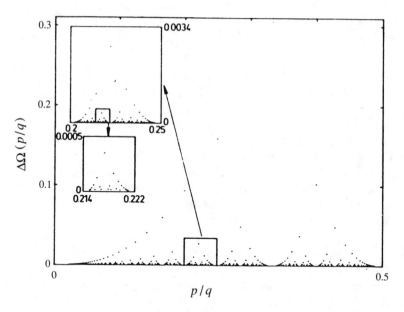

FIGURE 6.13   The width $\Delta\Omega(p/q)$ of the p/q mode-locked state as a function of the rational number p/q. From Jensen et al. (1984).

that determine θ_* and Ω_* are the equation for the fixed point and the equation for marginal stability.

$$f(\theta) = \theta_* = \theta_* + \Omega - \frac{k}{2\pi}\sin 2\pi\theta_*, \qquad (6.36)$$

$$\frac{df}{d\theta} = 1 = 1 - k\cos 2\pi\theta_*. \qquad (6.37)$$

Equation (6.37) implies that $\theta_* = \pm\frac{1}{4}$, which substituted into (6.36) gives $\Omega_* = \pm k/2\pi$, and thus in the interval $[0,1]$ we have $\Delta\Omega(0/1) = [0, k/2\pi)$.

Higher order resonances are similar. One seeks the solution (θ_*, Ω_*) to the set of equations

$$f_\Omega^q(\theta) = \theta + p, \qquad (6.38)$$

$$\frac{df_\Omega^q}{d\theta} = 1. \qquad (6.39)$$

An iterative technique based on Newton's method is described by Jensen et al. (1984) and explored in Exercise 6.2. The results are displayed in Figs. 6.13 and 6.14. Figure 6.13 shows the remarkable self-similar structure for the dependence of the widths of the mode-locked intervals on the rational number p/q that labels the mode-locked state. These widths were computed for $k = 1$. Figure 6.14 shows as a function of k the manner in which the rationals on the Ω-axis open out into finite widths as k increases. These intervals in the Ω-axis that expand out from the rational points are called *Arnold tongues* and are discussed in more detail in the next chapter.

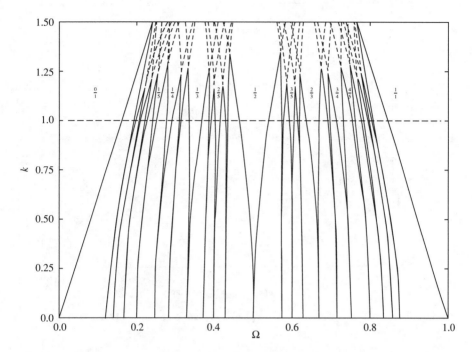

FIGURE 6.14   The development of finite widths for Ω in mode-locked states as k increases. From Jensen et al. (1984).

The winding number W as a function of Ω at $k = 1$ is given in Fig. 6.15. We see that at each value of Ω the winding number is some rational number. Such a plot is termed a *devil's staircase*. For $k = 1$ the measure of quasiperiodic states (Ω irrational) on the Ω-axis has become zero.

Let $S(\epsilon)$ denote the total width of all intervals larger than the given size ϵ. The space between the intervals is given by $1 - S(\epsilon)$, and the average number of gaps is defined by $N(\epsilon) = [1 - S(\epsilon)]/\epsilon$. Jensen et al. (1984) have shown that

$$N(\epsilon) \sim \left(\frac{1}{\epsilon}\right)^d \tag{6.40}$$

and find numerically that $d = 0.8700 \pm 3.7 \times 10^{-4}$. The computation of this result is analogous to the computation of the fractal dimension for the classical Cantor set in Section 4.2. From (6.40) we find

$$1 - S(\epsilon) = \epsilon^{1-d}, \tag{6.41}$$

and with $d \simeq 0.87$ we see that in the limit $\epsilon \to 0$ the total width of all intervals $S(\epsilon) \to 1$. The devil's staircase in Fig. 6.15 is then termed complete.

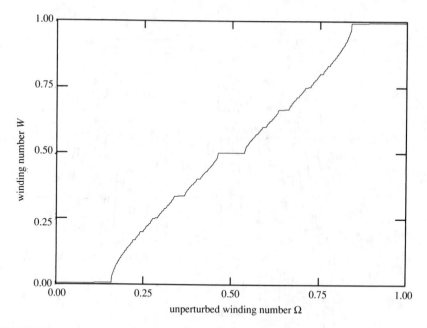

FIGURE 6.15   The devil's staircase plot of the winding number W vs. Ω.

6.6 Summary

The systems we have considered in this chapter have illustrated the richness of the chaotic behavior contained in the dynamics of nonlinear differential systems. To the extent that such nonlinear effects are ubiquitous in nature, can we expect detailed study to reveal chaos and fractal geometry. Even in the systems discussed in this chapter that have been studied for years, it is safe to say that all details have not yet been explained.

Exercises

6.1 For the differential system (6.15) of the disk dynamo, investigate the conditions for the flow to contract volume everywhere. Show that no spherical trapping region of the form (6.23) exists for this system.

6.2 (a) Write the two equations (6.38) and (6.39) in the fixed-point form

$$\mathbf{F}(\mathbf{x}) = 0, \qquad\qquad (6.42)$$

where $\mathbf{x} = (\theta, \Omega)$. Then verify by a Taylor expansion the Newton formula for iterative approximations to the solution of (6.42):

$$\mathbf{x}_{n+1} = \mathbf{x}_n - D\mathbf{F}(\mathbf{x}_n)^{-1} \cdot \mathbf{F}(\mathbf{x}_n). \qquad\qquad (6.43)$$

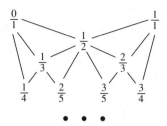

FIGURE 6.16 A Farey tree of rationals formed by combining rationals on the levels above according to $\frac{p}{q} \oplus \frac{r}{s} = \frac{p+r}{q+s}$.

(b) Show the recursion formulas for the sine circle map of (6.35):

$$\frac{\partial f_\Omega^{n+1}}{\partial \theta} = (1 - k \cos 2\pi\theta_n)\frac{\partial f_\Omega^n}{\partial \theta}, \tag{6.44}$$

$$\frac{\partial^2 f_\Omega^{n+1}}{\partial \Omega\, \partial \theta} = 2\pi k \sin 2\pi\theta_n \frac{\partial f_\Omega^n}{\partial \theta} \frac{\partial f_\Omega^n}{\partial \Omega} + (1 - k \cos 2\pi\theta_n)\frac{\partial^2 f_\Omega^n}{\partial \Omega\, \partial \theta}. \tag{6.45}$$

(c) Denote $\mathbf{x}_s$ as the superstable point where $df_{\Omega_s}^q/d\theta_s = 0$ and use Ω_s determined by $f_{\Omega_s}^q(0) = p$ as the starting point for a Newton iteration as described in parts (a) and (b). Find the widths of the first nine intervals for a Farey tree of rationals, as described in Fig. 6.16. Since the Arnold tongues are symmetric about $\Omega = \frac{1}{2}$ only five values need be computed.

(d) (For the energetic reader.) Compute enough additional intervals to verify the fractal dimension of the holes is $d \simeq 0.87$.

For results and suggestions see Jensen et al. (1984).

CHAPTER SEVEN

TWO-DIMENSIONAL MAPS

*Express the love of two lovers by a wedding of
variables, their mingling, their opposition, the
mysterious vibrations of kindred modes.*

We have seen in the previous chapter the important role played by
Poincaré maps in understanding the behavior of dynamical systems. These
Poincaré sections are two-dimensional maps, but usually it is not possible to
give the map explicitly. It is almost always necessary to integrate the flow and
to find the intersection points with the specified surface. But Poincaré maps
are not the only two-dimensional maps of interest. Certain dynamical systems
are directly modeled by two-dimensional maps and are important in their own
right. However, there does not seem to be any single two-dimensional map
that plays such a central role as the logistic map does in one dimension.
Consequently, we consider a variety of systems, chosen for their historical
importance or pedagogical value.

Many features of two-dimensional maps are similar to topics studied in
previous chapters. We encounter bifurcation behavior, normal forms, and
asymptotic sets. We study for the first time Arnold tongues and numerically
study strange attractors more closely.

7.1 Fixed Points

Denote a general two-dimensional map in the form

$$\mathbf{x}_{n+1} = \mathbf{f}(\mathbf{x}_n) \tag{7.1}$$

or more explicitly

$$\begin{aligned} x_{n+1} &= f(x_n, y_n), \\ y_{n+1} &= g(x_n, y_n), \end{aligned} \tag{7.2}$$

with a fixed point given by $x_{n+1} = x_n$ and $y_{n+1} = y_n$. We denote a fixed
point of (7.1) as $\mathbf{x}_*$, and usually we will choose coordinates so that $\mathbf{x}_*$ is at
the origin.

We do not have the convenient tool of graphical analysis for two-
dimensional maps, but just as with one-dimensional maps, the derivative of

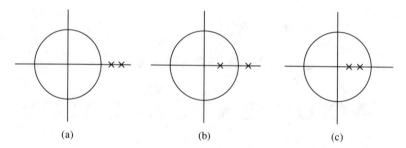

(a)   (b)   (c)

FIGURE 7.1   Three nondegenerate possibilities for real, positive eigenvalues λ_1, λ_2 of $D\mathbf{f}(\mathbf{x}_*)$.

the map usually determines the local stability. In this case the derivative is the matrix

$$D\mathbf{f}(\mathbf{x}_*) = \begin{pmatrix} f_{,x} & f_{,y} \\ g_{,x} & g_{,y} \end{pmatrix}\bigg|_{\mathbf{x}_*}, \tag{7.3}$$

where the comma notation denotes partial derivatives, i.e., $f_{,x} \equiv \partial f / \partial x$. Note that $D\mathbf{f}$ is evaluated at the fixed point $\mathbf{x}_*$.

Let us consider the difference vector

$$\delta_{n+1} = \mathbf{x}_{n+1} - \mathbf{x}_*$$

and consider the length of this vector

$$\|\delta_{n+1}\| = \|\mathbf{x}_{n+1} - \mathbf{x}_*\|, \tag{7.4}$$

where $\| \cdot \|$ denotes the usual Euclidean norm. Expanding in the usual way, we find

$$\|\delta_{n+1}\| \simeq \|D\mathbf{f}(\mathbf{x}_*) \cdot (\mathbf{x}_n - \mathbf{x}_*)\| = \|D\mathbf{f}(\mathbf{x}_*) \cdot \delta_n\|. \tag{7.5}$$

So it is clear that the length of δ_{n+1}, in comparison to δ_n, depends on what the matrix $D\mathbf{f}(\mathbf{x}_*)$ does to δ_n. For example the difference vector δ_{n+1} will be smaller in length, if the eigenvalues of $D\mathbf{f}(\mathbf{x}_*)$ have magnitude less than one. Thus we can classify the behavior at fixed points according to the eigenvalues of the matrix $D\mathbf{f}(\mathbf{x}_*)$.

If the eigenvalues are real, then the various possibilities for positive eigenvalues are depicted in Fig. 7.1 in relation to the unit circle in the complex plane. There are also obvious degenerate possibilities not pictured. The eigenvectors of $D\mathbf{f}(\mathbf{x}_*)$ give the linear manifolds tangent at $\mathbf{x}_*$ to the nonlinear invariant manifolds of $\mathbf{x}_*$. Locally, we can then choose coordinates such that

$$\mathbf{x}_1 = (x_1, y_1) = (\lambda_1 x_0, \lambda_2 y_0). \tag{7.6}$$

Hence under iterations of the map $\mathbf{x}_n = (\lambda_1^n x_0, \lambda_2^n y_0)$. Eliminating n,

$$\frac{\log|x_n|}{\log \lambda_1} - \frac{\log|y_n|}{\log \lambda_2} = \text{constant}. \tag{7.7}$$

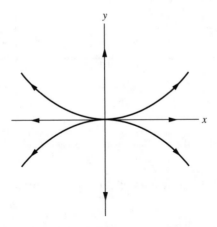

FIGURE 7.2   Sketch of the invariant manifolds (x and y axes) and curves for case (a) of Fig. 7.1. For case (c) all arrows would be reversed.

For $\lambda_1 > 1$, $\lambda_2 > 1$, and $\lambda_1 \neq \lambda_2$ the points are on an orbit such as depicted in Fig. 7.2. For $\lambda_1 < 1$, and $\lambda_2 < 1$ the sketch is qualitatively similar to Fig. 7.2, but with the direction of all arrows reversed. A fixed point with eigenvalues of $D\mathbf{f}(\mathbf{x}_*)$ as in Fig. 7.1a is called an *unstable node* and case (c) is called a *stable node*.

If $\lambda_1 > 1$ and $\lambda_2 < 1$, then the invariant manifolds and orbits under the map are as sketched in Fig. 7.3. Such a fixed point is called a *saddle* and is also unstable.

Since $\mathbf{f}$ is real valued, if the eigenvalues are complex, then we can let $\lambda_1 \equiv \lambda = ae^{i\alpha}$ and $\lambda_2 = \lambda^*$, where $0 < \alpha < \pi$. Figure 7.4 sketches the possible positions of eigenvalues in the complex plane with respect to the unit circle.

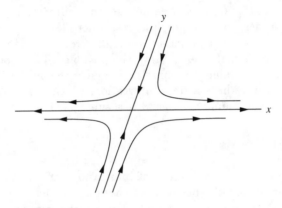

FIGURE 7.3   Sketch of invariant manifolds and curves for a saddle fixed point.

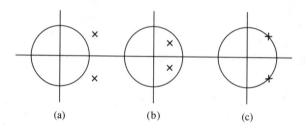

FIGURE 7.4   Sketch of the possible positions of eigenvalues of $D\mathbf{f}(\mathbf{x}_*)$ in the complex plane.

With $\lambda = ae^{i\alpha}$ the linear part of the map can be put in the form

$$D\mathbf{f}(\mathbf{x}_*) = a \begin{pmatrix} \cos \alpha & -\sin \alpha \\ \sin \alpha & \cos \alpha \end{pmatrix}, \tag{7.8}$$

and then locally the map looks like

$$r_{n+1} = ar_n, \qquad \theta_{n+1} = \theta_n + \alpha. \tag{7.9}$$

We can again consider n applications of the map from initial values (r_0, θ_0) to find $r_n = a^n r_0$ and $\theta_n = \theta_0 + n\alpha$. Eliminating n we find that the invariant curves of this map are given by $r \exp(-\theta \log a/\alpha) =$ constant. If $a \neq 1$ these local invariant curves are logrithmic spirals. Figure 7.5 sketches such an invariant curve in the neighborhood of the fixed point $\mathbf{x}_*$. The figure is drawn for $a > 1$. The direction of the arrows is reversed in Fig. 7.5 if $a < 1$. A fixed point $\mathbf{x}_*$ of this type with complex eigenvalues for $D\mathbf{f}(\mathbf{x}_*)$ is called a *focus* and is unstable if $a > 1$, as in Fig. 7.4a. For $a < 1$, as in Fig. 7.4b, the fixed point is a stable focus.

    If the eigenvalues are not on the unit circle, then the linear behavior gives a reliable picture of the stability properties of the fixed point $\mathbf{x}_*$. Just as with one-dimensional maps, or the differential flows examined in Chapter 5, these two-dimensional maps often depend on a bifurcation parameter that we call μ. Consider, for instance, the case where $|\lambda_2|$ remains < 1 but λ_1 passes through 1 or -1 as μ passes through zero. Then the map rapidly becomes one-dimensional, and we have the possible bifurcations as listed in Table 2.2. If the eigenvalues are complex and cross the unit circle as μ passes through zero, then we have a Hopf bifurcation, which has no counterpart in one-dimensional maps and is discussed in a subsequent section.

7.2 Normal Forms

In order to understand the nature of bifurcations at critical values of the map parameter μ, and the behavior of the map on the center manifold of a fixed point, we undertake a study of normal forms similar in spirit to our

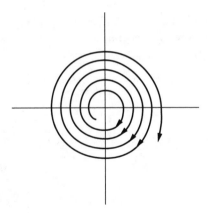

FIGURE 7.5 Sketch of an unstable focus. A stable focus has the direction of the arrows reversed.

study for differential flows. The end goal of the analysis is exactly the same as Section 5.5, namely to push the nonlinear behavior to as high an order as possible through coordinate transformations. The polynomial forms that remain after this process are referred to as normal forms. This is entirely analogous to the situation for differential flows as discussed in Section 5.5.

For two-dimensional maps the use of complex coordinates is convenient. This is primarily because the eigenvectors and eigenvalues of the maps can be complex. Indeed, because cases with real eigenvalues reduce to essentially one-dimensional maps except for degeneracy, we focus in this section on the complex case.

By choosing $f(x_n, y_n)$ to be a new y coordinate, we can always write the map (7.2) in the form

$$
\begin{aligned}
x_{n+1} &= y_n, \\
y_{n+1} &= Ax_n + By_n + \sum_{\substack{j,k \\ j+k \geq 2}} g_{jk} x_n^j y_n^k,
\end{aligned}
\tag{7.10}
$$

where these coordinates have further been chosen so that the fixed point is at the origin. From the linear part of (7.10) we see that $A = -\lambda\lambda^*$ and $B = \lambda + \lambda^*$. We take the eigenvalue λ to be of the form

$$
\lambda = (1 + \mu)e^{i\alpha},
\tag{7.11}
$$

with μ a parameter. The complex pair of eigenvalues cross the unit circle along the ray specified by the angle α, $0 < \alpha < \pi$, when μ passes through zero.

The eigenvectors of $D\mathbf{f}(0)$ are proportional to

$$
\begin{pmatrix} 1 \\ \lambda \end{pmatrix}, \quad \begin{pmatrix} 1 \\ \lambda^* \end{pmatrix}.
\tag{7.12}
$$

We choose complex coordinates z and z^* so that one iteration of the map corresponds in linear order to multiplying by λ or λ^*, respectively. Thus the transformation to complex coordinates is given by

$$\begin{pmatrix} x \\ y \end{pmatrix} = c \begin{pmatrix} 1 & 1 \\ \lambda & \lambda^* \end{pmatrix} \begin{pmatrix} z \\ z^* \end{pmatrix}, \tag{7.13}$$

where $c^{-1} = -i(\lambda - \lambda^*)$. Consequently, from (7.13) we obtain the transformation

$$z = i\lambda^* x - iy, \qquad z^* = -i\lambda x + iy, \tag{7.14}$$

$$x = \frac{i}{\lambda - \lambda^*}(z + z^*), \qquad y = \frac{i}{\lambda - \lambda^*}(\lambda z + \lambda^* z^*). \tag{7.15}$$

Using (7.10) we then obtain for the map

$$z_{n+1} = \lambda z_n + \sum_{\substack{jk \\ j+k \geq 2}} \bar{g}_{jk} z_n^j z_n^{*k}, \tag{7.16}$$

plus its complex conjugate. The $\bar{g}_{jk}$ are linear combinations of the g_{jk} with polynomials in the eigenvalues λ and λ^* comprising the coefficients.

As in Section 5.5, we now make a coordinate transformation

$$w = z + \sum_{\substack{j,k \\ j+k \geq 2}} p_{jk} z^j z^{*k} \tag{7.17}$$

and choose the coefficients p_{jk} to eliminate as many nonlinear terms as possible, beginning with quadratic terms. Proceeding in this way,

$$\begin{aligned} w_{n+1} &= z_{n+1} + \sum_{\substack{j,k \\ j+k \geq 2}} p_{jk} z_{n+1}^k z_{n+1}^{*k} \\ &= \lambda w_n + \sum_{\substack{j,k \\ j+k \geq 2}} b_{jk} w_n^j w_n^{*k}. \end{aligned} \tag{7.18}$$

In Eq. (7.18) we substitute from (7.16) for z_{n+1} and from (7.17) for w_n and similarly from the conjugate equations for z_{n+1}^* and w_n^*. Then we equate the coefficients of the various monomials of the form $z_n^j z_n^{*k}$ on both sides of the equation. The algebra can be tedious, and we only give the equations up through third order.

$$z_n^2 : b_{20} + \lambda p_{20}(1 - \lambda) = \bar{g}_{20}. \tag{7.19a}$$

$$z_n z_n^* : b_{11} + \lambda p_{11}(1 - \lambda^*) = \bar{g}_{11}. \tag{7.19b}$$

$$z_n^{*2} : b_{02} + p_{02}(\lambda - \lambda^{*2}) = \bar{g}_{02}. \tag{7.19c}$$

$$z_n^3 : b_{30} + \lambda p_{30}(1 - \lambda^2) = -p_{02}^* b_{11} - 2b_{20}p_{20}$$
$$+ \lambda(2\bar{g}_{20}p_{20} + \bar{g}_{02}^* p_{11}) + \bar{g}_{30}. \tag{7.19d}$$

$$z_n^2 z_n^* : b_{21} + \lambda p_{21}(1 - \lambda\lambda^*) = -(p_{11}^* + p_{20})b_{11} - 2p_{02}^* b_{02} - 2p_{11}b_{20}$$
$$+ \lambda^*(p_{11}\bar{g}_{20} + 2p_{20}\bar{g}_{02}^*) + \lambda(2p_{20}\bar{g}_{11} + p_{11}\bar{g}_{11}^*) + \bar{g}_{21}. \tag{7.19e}$$

$$z_n z_n^{*2} : b_{12} + \lambda p_{12}(1 - \lambda^{*2}) = -(p_{20}^* + p_{11})b_{11} - 2p_{11}^* b_{02} - 2p_{02}b_{20}$$
$$+ \lambda^*(p_{11}\bar{g}_{11} + 2p_{02}\bar{g}_{11}^*) + \lambda(2p_{20}\bar{g}_{02} + p_{11}\bar{g}_{20}^*) + \bar{g}_{12}. \tag{7.19f}$$

$$z_n^{*3} : b_{03} + p_{03}(\lambda - \lambda^{*3}) = -2p_{20}^* b_{02} - p_{02}b_{11}$$
$$+ \lambda^*(2p_{02}\bar{g}_{20}^* + p_{11}\bar{g}_{02}) + \bar{g}_{03}. \tag{7.19g}$$

The strategy is to choose the p_{ij} in Eqs. (7.19) so that the $b_{ij} = 0$. This is possible only so long as the coefficient of the relevant p_{ij} is not equal to zero. These coefficients are always polynomials in λ and λ^* and need only be examined on the unit circle, i.e., for $\mu = 0$. We take $0 < \alpha < \pi$ and let $\alpha = 2\pi l/m$. We refer to a value of α for which a b_{ij} cannot be set equal to zero as a *resonance*. These resonances are analogous to the mode-locked states for circle maps. They are manifest in the two-dimensional mapping by a finite number of image points under iteration of the map.

From Eqs. (7.19) it is clear that b_{20} and b_{11} can always be chosen equal to zero. For an $l : m = 1 : 3$ resonance b_{02} cannot be made zero, and (7.18) will have in it a term of the form w_n^{*2}. Remember that we consider the case for loss of linear stability where $\mu = 0$ and that the 1:3 resonance corresponds to $\lambda^3 = 1$. From (7.19e) we see that the term $w_n^2 w_n^*$ is always in resonance, i.e., choosing p_{21} can never make b_{21} vanish. The 1:4 resonance leads to a term w_n^{*3} in the normal form.

Assuming that we do not have a 1:3 resonance we find from (7.19a–c)

$$p_{20} = \frac{\bar{g}_{20}}{\lambda(1 - \lambda)}, \quad p_{11} = \frac{\bar{g}_{11}}{\lambda(1 - \lambda^*)}, \quad p_{02} = \frac{\bar{g}_{02}}{\lambda - \lambda^{*2}}, \tag{7.20}$$

and

$$b_{21} = \frac{|\bar{g}_{11}|^2}{1 - \lambda^*} + \frac{2|\bar{g}_{02}|^2}{\lambda^2 - \lambda^*} + \frac{2\lambda - 1}{\lambda(1 - \lambda)}\bar{g}_{11}\bar{g}_{20} + \bar{g}_{21}. \tag{7.21}$$

For completeness we also note

$$\bar{g}_{20} = -\bar{g}_{02}^* = i(g_{02}\lambda^2 + g_{11}\lambda + g_{20})/(\lambda^* - \lambda)^2,$$

$$\bar{g}_{11} = 2i(g_{02} + g_{20} + g_{11}\text{Re}(\lambda))/(\lambda^* - \lambda)^2,$$

$$\bar{g}_{30} = -\bar{g}_{03}^* = (g_{03}\lambda^3 + g_{12}\lambda^2 + g_{21}\lambda + g_{30})/(\lambda^* - \lambda)^3,$$

$$\bar{g}_{21} = -\bar{g}_{12}^* = (3g_{03}\lambda + g_{12}(2 + \lambda^2) + g_{21}(2\lambda + \lambda^*) + 3g_{30})/(\lambda^* - \lambda)^3.$$

7.3 Hopf Bifurcation and Arnold Tongues

To consider the simplest case first, we assume the map has no resonances and hence can be put in the normal form

$$z_{n+1} = \lambda z_n + b_{21}z_n^2 z_n^*, \tag{7.22}$$

with higher order terms ignored. Changing to polar coordinates, $z = re^{i\theta}$, (7.22) becomes

$$r_{n+1} = \left|(1 + \mu)r_n e^{i\alpha} + b_{21}r_n^3\right|. \tag{7.23}$$

For convenience we write $b_{21} = -Be^{i\beta}$, where $B > 0$. Substituting into (7.23) leads to the result

$$r_{n+1}^2 = r_n^2(1 + 2\mu) - 2Br_n^4(1 + \mu)\cos(\beta - \alpha) + B^2 r_n^6. \tag{7.24}$$

We can regard this as a one-dimensional map in r, and assuming that the B^2 term can be discarded, we find a fixed point of (7.24), which is called the Hopf radius.

$$r_H^2 \simeq \frac{\mu}{B\cos(\beta - \alpha)}. \tag{7.25}$$

If $\cos(\beta - \alpha) > 0$, then for $\mu > 0$ we have a Hopf circle of radius r_H [ellipse in the original (x, y) coordinates]. The stability is examined by letting $r_n = r_H + \delta$, with δ assumed small.

$$r_n^2 \simeq r_H^2 + 2r_H\delta. \tag{7.26}$$

Substituting into (7.24) we find

$$r_{n+1}^2 \simeq r_H^2 + 2r_H(\delta - 2r_H^2 B\cos(\beta - \alpha)), \tag{7.27}$$

and thus the deviation in r_{n+1} from r_H will be smaller than δ. This bifurcation is depicted in Fig. 7.6a and is referred to as a *supercritical* Hopf bifurcation.

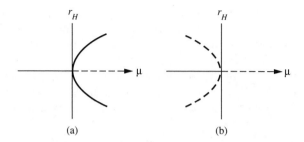

(a)   (b)

FIGURE 7.6   (a) A supercritical Hopf bifurcation and (b) a subcritical Hopf bifurcation.

If $\cos(\beta - \alpha) < 0$, then there exists a Hopf circle for $\mu < 0$, with radius given by (7.25). Equation (7.27) shows this Hopf circle to be unstable. This bifurcation is depicted in Fig. 7.6b and is called *subcritical*. The nonlinear terms do not stabilize the mapping.

If $\cos(\beta - \alpha) \approx 0$, then the term of the order r_n^6 is important for determining r_{n+1}. To obtain the correct coefficient for r_n^6, it may be necessary to keep additional terms in the normal form (7.22). In any case there results an equation analogous to (7.24):

$$r_{n+1}^2 = r_n^2(1 + 2\mu) - 2B\cos(\beta - \alpha)r_n^4 + ar_n^6, \tag{7.28}$$

where the coefficient a now denotes that other terms in (7.22) are perhaps necessary for (7.28) to be correctly ordered. For the fixed-point radii we find

$$R_\pm^2 = \left[B\cos(\beta - \alpha) \pm \sqrt{B^2\cos^2(\beta - \alpha) - 2\mu a}\right]/a. \tag{7.29}$$

The quantities B, and β are determined by (7.21), and a is similarly determined by the normal form construction for the map.

With $a > 0$, $B\cos(\beta - \alpha) > 0$, and $\mu \simeq 0$ we find two Hopf radii R_+ and R_-, and also find that R_- is stable and R_+ is unstable. The bifurcation diagram is sketched in Fig. 7.7.

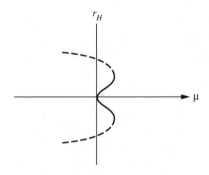

FIGURE 7.7   Singular type of Hopf bifurcation called a crater bifurcation.

To see the emergence of Arnold tongues, we specifically consider the case of resonance. The resonances occur when the eigenvalue λ is a root of unity, i.e., $\lambda^m = 1$. By the restriction $0 < \alpha < \pi$, we force $m \geq 3$. The cases $m = 3$ and $m = 4$, i.e., the resonances 1:3 and 1:4, are termed *strong resonances*. Resonances for $m \geq 5$ are called *weak resonances*. We note from (7.19) that a resonance of order m implies a term in the normal form for the map of the form z^{*m-1}. Consequently, we assume a two-dimensional map of the form

$$z_{n+1} = \lambda z_n + b_{21} z_n^2 z_n^* + b_{m-1} z^{*m-1}, \qquad (7.30)$$

where we have dropped the higher order terms for simplicity. We restrict our attention to cases of weak resonance. Furthermore we write the eigenvalue λ in a slightly more general form

$$\lambda = (1 + \mu e^{i\phi}) e^{i\alpha}, \qquad (7.31)$$

which leads to the generalization

$$r_H^2 = \frac{\mu \cos \phi}{B \cos(\beta - \alpha)} \qquad (7.32)$$

for the Hopf radius, again to lowest order. Equation (7.32) is a slight generalization of (7.25).

At the Hopf radius we obtain from (7.30) that

$$\theta_{n+1} = \theta_n + \alpha + \arg\left(1 + \mu e^{i\phi} + r_H^2 b_{21} e^{-i\alpha} + \bar{C} r_H^{m-2} e^{-im\theta_n}\right), \qquad (7.33)$$

where $\bar{C}$ is a complex constant. Substitution of (7.32) results in

$$\theta_{n+1} = \theta_n + \alpha + \arg\left(1 + \frac{i\mu \sin(\phi - \beta + \alpha)}{\cos(\beta - \alpha)} + \bar{C} \mu^{\frac{m-2}{2}} e^{-im\theta_n}\right). \qquad (7.34)$$

We recall that we are assuming a resonance of the form $\alpha = 2\pi l/m$, and to have an m-period cycle we need only have the imaginary part of the quantity in the parentheses in (7.34) vanish, i.e.,

$$\frac{\mu \sin(\phi - \beta + \alpha)}{\cos(\beta - \alpha)} + \mu^{\frac{m-2}{2}} \text{Im}(\bar{C} e^{-im\theta}) = 0. \qquad (7.35)$$

This is possible provided an inequality of the form

$$|\sin(\phi - \beta + \alpha)| \leq \tilde{C} \mu^{\frac{m-4}{2}} \qquad (7.36)$$

is satisfied. The actual form of the constant $\tilde{C}$ is determined by $\bar{C}$, which is in turn determined by the coefficients in the normal form for the map. The actual value of the constant $\tilde{C}$ is for our present purposes not important. It

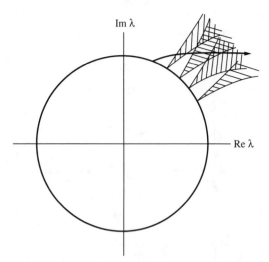

FIGURE 7.8   For forward Hopf bifurcation a few Arnold tongues corresponding to resonances $\alpha/2\pi = l/m$ are sketched. Nonoverlapping regions correspond to mode-locking. Overlap indicates a parameter region for chaos. A possible parameter path through such regions is indicated.

is important to note as μ increases that values of ϕ centered about $\alpha - \beta$ and contained between two arcs satisfy (7.36). Such a region is referred to as an *Arnold tonque* after V.I. Arnold who elucidated the role of such resonances (Arnold, 1977; 1983, Section 25). Every weak resonance occurring in the normal form for such a two-dimensional map gives rise to such an Arnold tonque. Figure 7.8 depicts several such Arnold tonques emanating out (forward Hopf bifurcation) from the unit circle. Compare Fig. 7.8 with Fig. 6.16, the previous example of Arnold tongues in the sine circle map. A possible parameter path through such tonques as might result from a second parameter dependence is also sketched. Moving along the parameter path sketch in Fig. 7.8, the system encounters Arnold tongues corresponding to regions of a specific resonance (mode-locking) followed by regions of resonance overlap where we might expect chaotic behavior. The cases of strong resonance and subtleties such as $\cos(\beta - \alpha) \approx 0$ have been omitted from our introductory discussion of two-dimensional maps. For further details we refer the interested reader to the references given previously.

We close our discussion of normal forms by considering an example discussed in Lauwerier (1986, p.74). Consider the map

$$
\begin{aligned}
x_{n+1} &= ax_n(1 - x_n - y_n), \\
y_{n+1} &= bx_n y_n,
\end{aligned}
\tag{7.37}
$$

where the parameters satisfy $2 < a, b \leq 4$. Figure 7.9 shows the effect of one application of the map (7.37) to a set of points on the unit circle. This map has three fixed points $(0,0)$, $(1 - 1/a, 0)$, and $(1/b, 1 - 1/b - 1/a)$. We focus our attention on the last of these. As usual, we determine stability of this fixed point by looking at the linearization Df of (7.37) and its eigenvalues.

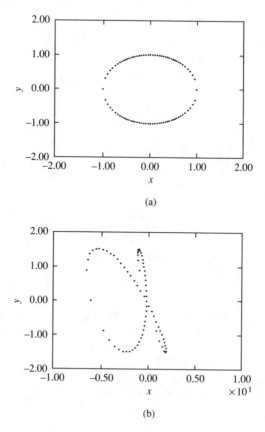

FIGURE 7.9   To a circle of points (a) the map (7.37) is applied once resulting in (b).

The eigenvalues are given by

$$\lambda_{\pm} = \frac{1}{2}\left[(2 - a/b) \pm \sqrt{(2 - a/b)^2 - 4a(1 - 2/b)}\right]. \qquad (7.38)$$

The stability regions are sketched in Fig. 7.10. Figure 7.10 also shows two paths in parameter space. There is nothing special about these curves; they serve simply as a convenient choice to demonstrate numerically the loss of stability for the selected fixed point and the appearance of resonances and chaos. The corresponding curves for the eigenvalues are given in Fig. 7.11.

The boundary curve between the stable and unstable regions in Fig. 7.10 is given by $b = 2a/(a - 1)$. Along this boundary we have

$$\lambda_{\pm} = \frac{5 - a}{4} \pm \frac{i}{4}\sqrt{(a - 1)(9 - a)} \qquad (7.39)$$

giving $\cos \alpha = (5 - a)/4$. The parameter μ is given by $\mu = -1 + \sqrt{a - 2a/b}$.

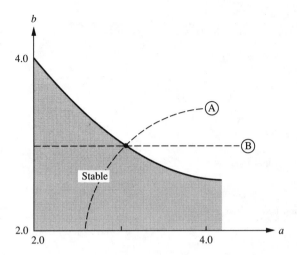

FIGURE 7.10   Stability regions in parameter space for the two-dimensional map of (7.37). The curves labeled A and B denote two curves in this parameter space with corresponding curves sketched in eigenvalue space in Fig. 7.11.

At $a = 3$ there is a weak 1:6 resonance with $\alpha = \pi/3$. Equation (7.21) gives $b_{21} = -3i\sqrt{3}$ and consequently $\beta = \pi/2$. There is an Arnold tongue centered on the ray $\beta - \alpha = \pi/6$ and a Hopf radius given by $r_H^2 = 2\mu\cos\phi/9$.

The curve labeled B in Fig. 7.11 intersects the Arnold tongue emanating out of the unit circle from $\alpha = 2\pi/5$, i.e., the 1:5 resonance. This is easily seen numerically as one examines the attracting sets for parameter values

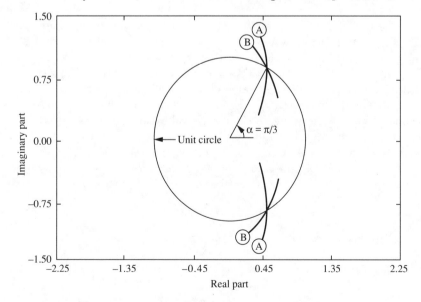

FIGURE 7.11   Eigenvalue curves corresponding to the parameter curves in Fig. 7.10. Both curves cross the unit circle at $\alpha=\pi/3$ when both parameters equal 3.0.

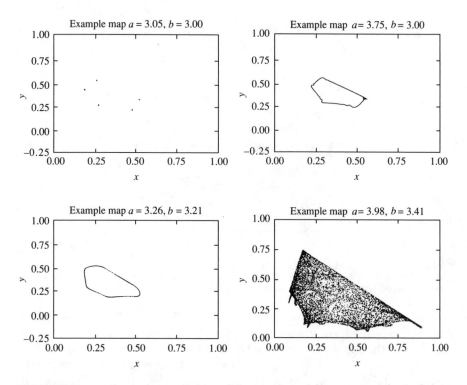

FIGURE 7.12   Attracting sets for four different values of the parameters located along the curves A and B in Fig. 7.10.

along the curve B. The map (7.37) is extremely easy to study numerically, and the reader is strongly urged to do some numerical experimentation by exploring the evolution of the attracting set as the parameter values change along one of the curves A or B. Some examples of the attracting sets [initial point $(0.2,0.3)$] for assorted parameter values along the curves A and B in Fig. 7.11 are shown in Fig. 7.12. As one proceeds along the curve B in parameter space, depending on the precise values of the parameters, one may observe first a fixed point, followed by a $1:5$ resonance, then a Hopf circle, and finally chaos. Some of these attractor sets are depicted in Fig. 7.12 where an initial transient of 100 points has been discarded. The capacity fractal dimension of the strange attractor shown in the last figure is approximately 1.98.

7.4 The Hénon Map

The Hénon map (Hénon, 1976)

$$
\begin{aligned}
x_{n+1} &= 1 - ax_n^2 + y_n, \\
y_{n+1} &= bx_n
\end{aligned}
\tag{7.40}
$$

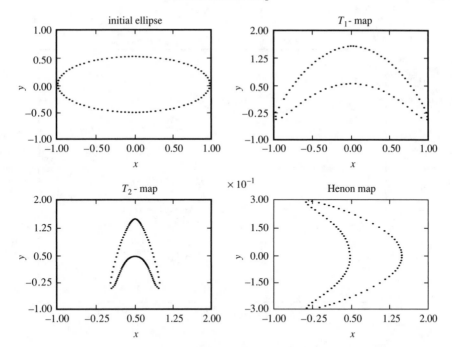

FIGURE 7.13   The successive application of the maps comprising the Hénon map to an ellipse of points.

has been studied extensively and is, in a sense, the simplest extension of the logistic map to two-dimensional mappings. Part of the reason for this is because the Hénon map may be decomposed into three separate maps. We denote the map of (7.40) as $T: \mathbb{R}^2 \to \mathbb{R}^2$ and write it in the form $T = T_3 \circ T_2 \circ T_1$. The map T_1, $(x, y) \mapsto (x, 1 - ax^2 + y)$ is an area-preserving, nonlinear bending. The map T_2, $(x, y) \mapsto (bx, y)$ is a contraction in the x direction. The map T_3, $(x, y) \mapsto (y, x)$ maps the bent and contracted area back onto itself. For the parameter values chosen by Hénon, $a = 1.4$, $b = 0.3$, the effect of this sequence of maps on an initial ellipse of points is shown in Fig. 7.13. The effect of this repeated bending and contracting is to produce an attracting set that is a strange attractor.

Figure 7.14a shows 1000 points in the attracting set of the initial point (0.63135448,0.18940634) after having discarded an initial transient of 100 points. The contracting properties are determined solely by the parameter b, since $\det(Df) = -b$. With $b = 0.3$ the contraction in one iteration is mild enough that the sheaves of the strange attractor are visible. This is in contrast to the contraction in the Lorenz map, as discussed in Section 6.3. Figure 7.14 is a series of plots in the manner of Hénon that shows increasingly greater magnification of a small region on the strange attractor. The self-similar and infinitely nested nature of the points in this part of the attractor is evident. To obtain the 1000 points in Fig. 7.14d, i.e., the window in Fig. 7.14c, required more than 10^7 iterations of the Hénon map.

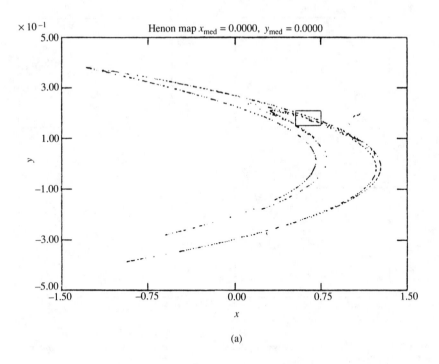

(a)

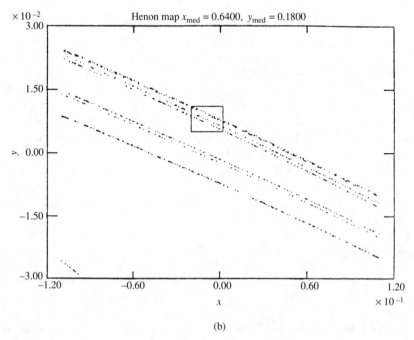

(b)

FIGURE 7.14   The strange attractor of the Hénon map (7.40) with $a=1.4$ and $b=0.3$. The small window regions in (a)–(c) are magnified in the subsequent plots (b)–(d), respectively. Each plot contains 1000 points.

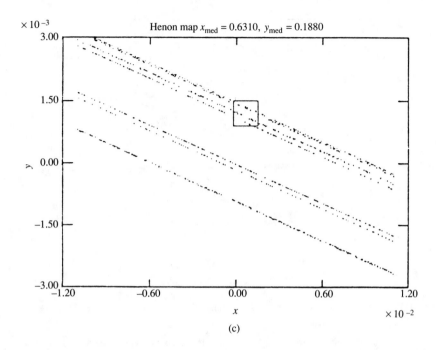

(c)

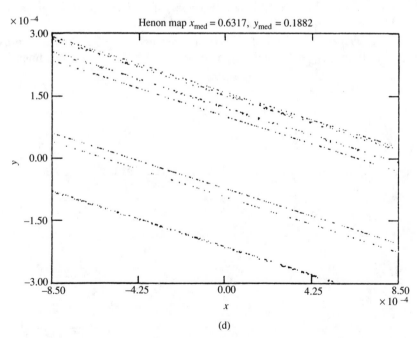

(d)

FIGURE 7.14c,d

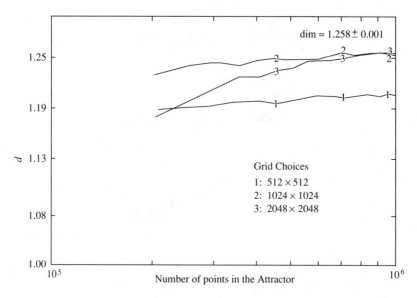

FIGURE 7.15   Convergence of the capacity dimension of the Hénon attractor.

Note that the Hénon map is invertible, in contrast to the map in the previous section. Noninvertibility is necessary for chaos in one-dimensional maps but not for maps on higher dimensional spaces.

The capacity dimension of the strange attractor in Fig. 7.14 is approximately 1.26 and Fig. 7.15 shows the convergence of the capacity dimension

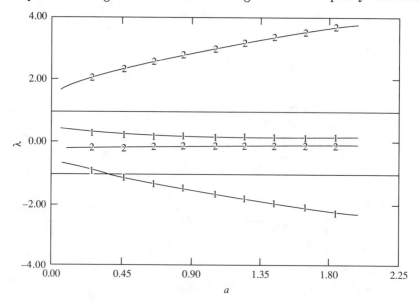

FIGURE 7.16   The eigenvalue tracks for the fixed points x_1 and x_2 as a function of the parameters a and b in the Hénon map (7.40).

calculation as $\epsilon \to 0$ to this value. The Hénon attractor has been one of the classical examples examined by authors in the calculation of fractal dimension.

The behavior of this map as the parameters a and b are varied may also be studied. Devaney (1987, Section 2.9) in a long series of exercises studies the bifurcation properties of the Hénon map. The fixed points of (7.40) are easily found to be

$$x_{1,2} = \left[b - 1 \pm \sqrt{b^2 - 2b + 4a + 1}\right]/2a, \tag{7.41}$$

and the eigenvalues of Df at these fixed points are

$$\lambda = -ax_* \pm \sqrt{a^2 x_*^2 + b}, \tag{7.42}$$

where x_* is either fixed-point coordinate x_1, x_2.

Figure 7.16 shows the track of the eigenvalues as the parameter a is varied while holding $b = 0.3$. In the range of parameters considered the eigenvalues are real. Furthermore, since $\det(Df) = -b = \lambda_+\lambda_-$, for the case of $b = 0.3$, a Hopf bifurcation is not an issue. However, from Fig. 7.16 we see that x_2 is an unstable saddle point and that x_1 becomes unstable at $a \approx 0.37$. Derrida et al. (1979) have shown $a = 0.3675$ to be the beginning of a period-doubling transistion to chaos with $a_\infty \simeq 1.058048$.

7.5 Renormalization of an Area-Preserving Map

The period-doubling transistion to chaos mentioned at the conclusion of the previous section suggests that we might try a renormalization calculation similar to that in Section 3.1. We consider a renormalization calculation of the Hénon map (7.40) for the area-preserving case where $b = -1$. Area-preserving maps are special because the whole issue of attracting sets vanishes for such maps. There are no attracting fixed points, attracting limit cycles, or strange attractors. We consider area-preserving dynamics in detail in the next chapter.

We consider

$$\begin{aligned} x_{n+1} &= 1 - ax_n^2 + y_n, \\ y_{n+1} &= -x_n. \end{aligned} \tag{7.43}$$

It is straightforward to compute the fixed points of this map and then to move the origin of coordinates to the smaller of the fixed points. With an appropriate rescaling of the coordinates, the map (7.43) can be written in the form

$$\begin{aligned} x_{n+1} &= 2cx_n + 2x_n^2 - y_n, \\ y_{n+1} &= x_n. \end{aligned} \tag{7.44}$$

The details of this transformation are the subject of Exercise 7.5.

The fixed points of (7.44) are $(x_*, y_*) = (0,0), (1 - c, 1 - c)$. The origin is stable for $|c| < 1$. If we denote the map of (7.44) as $F: \mathbb{R}^2 \to \mathbb{R}^2$. Then the fixed points of $F^2 \equiv F \circ F$ (2-cycle points of F) are given by $(\bar{x}_n, \bar{x}_{n+1})$ and $(\bar{x}_{n+1}, \bar{x}_n)$, where

$$\bar{x}_n = \frac{1}{2}\left[-(c+1) + (-1)^n \sqrt{(c+1)(c-3)}\right], \quad n = 0, 1, 2, \ldots \quad (7.45)$$

In the manner of Section 3.1, we transform the origin of F^2 to one of the 2-cycle points. Specifically, we let $x_n = \bar{x}_n + \xi_n$ and $y_n = \bar{x}_{n+1} + \eta_n$ and find, keeping terms only up through quadratic order

$$\begin{aligned}
\xi_{n+2} &= \xi_n(A_n A_{n+1} - 1) - A_{n+1}\eta_n + 2\eta_n^2 \\
&\quad + \xi_n^2(2A_n^2 + 2A_{n+1}) - 4A_n\eta_n\xi_n, \quad (7.46) \\
\eta_{n+2} &= A_n\xi_n + 2\xi_n^2 - \eta_n, \quad (7.47)
\end{aligned}$$

where $A_n = 2c + 4\bar{x}_n$ and $A_{n+1} = 2c + 4\bar{x}_{n+1}$.

Clearly (7.46) and (7.47) are not of the form (7.44). However, for this map it is not difficult to achieve this form. Let

$$\chi_n = A_n\xi_n + 2\xi_n^2 - \eta_n, \quad (7.48)$$

and then (7.47) becomes

$$\eta_{n+2} = \chi_n, \quad (7.49)$$

which gets us the second of (7.44). Using (7.46) we find

$$\begin{aligned}
\chi_{n+2} &= (A_n A_{n+1} - 2)\chi_n - \eta_n - \frac{4}{A_n^2}(A_n A_{n+1} - 2)\chi_n\eta_n + \frac{4}{A_n^2}\eta_n^2 \\
&\quad + \chi_n^2\left[\frac{2}{A_n}(A_n^2 + A_{n+1}) + \frac{2}{A_n^2}(A_n A_{n+1} - 1)(A_n A_{n+1} - 2)\right]. \quad (7.50)
\end{aligned}$$

The unwanted quadratic terms η_n^2 and $\chi_n\eta_n$ can further be eliminated without altering (7.49) by letting $u_n = \chi_n - 2\chi_n^2/A_n^2$ and $v_n = \eta_n - 2\eta_n^2/A_n^2$. Equations (7.49) and (7.50) assume the form

$$\begin{aligned}
u_{n+2} &= 2c'u_n + 2\alpha u_n^2 - v_n, \\
v_{n+2} &= u_n. \quad (7.51)
\end{aligned}$$

Figure 7.17 shows the effect of the maps F^2, F and the renormalization transformation. Consequently, the new value of the map parameter is

$$c' = \frac{1}{2}(A_n A_{n+1} - 2) = 2(c + 2\bar{x}_n)(c + 2\bar{x}_{n+1}) - 1 = -2c^2 + 4c + 7. \quad (7.52)$$

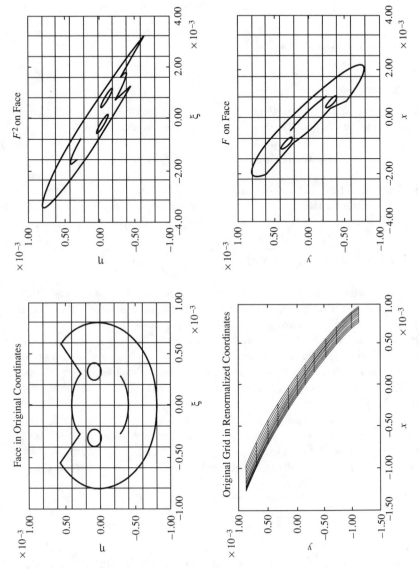

FIGURE 7.17 Sketch illustrating F^2, F and the renormalization transformation.

We approximate the accumulation value of the map parameter c for the period-doubling sequence by setting $c' = c$ in (7.52) and then solving for the value of c:

$$c_\infty \simeq -1.2656$$

Also as was done in Section 3.1, dc'/dc computed from (7.52) and evaluated at c_∞ gives an analytic approximation to δ:

$$\delta \simeq 9.0624.$$

Many authors, including Helleman (1980) and Greene et al. (1981), report the numerical determination of these quantities, namely

$$c_\infty = -1.266311\ldots, \qquad \delta = 8.721097\ldots$$

It is clear that the straightforward renormalization procedure gives values remarkably close to the more accurate values determined numerically. For a calculation similar to that in this section see Hu (1982).

7.6 Universality for Area-Preserving Maps

The renormalization calculation of the previous section has many similarities with the renormalization calculation for one-dimensional maps outlined in Section 3.1. Just as this renormalization for one-dimensional maps leads to universal features for period-doubling sequences in one-dimensional maps, there is a universality for area-preserving maps of the plane undergoing period-doubling sequences to chaos. Our main reference for this section is Greene et al. (1981), however, essentially the same results have been obtained by Collet et al. (1981).

A complete discussion of the classification of bifurcations and periodic points for area-preserving, two-dimensional maps requires a more lengthy treatment than space permits in the present work. Therefore we restrict ourselves to a discussion of the main results and refer the interested reader to the references cited for more detail.

As in the previous sections, we refer to the map on the plane as

$$F: I\!R^2 \to I\!R^2. \tag{7.53}$$

Greene et al. (1981) focus their attention on maps having a symmetry. A symmetry is defined as a map on the plane $S: I\!R^2 \to I\!R^2$ such that S and $F \circ S$ are reflections. A *reflection* $R: I\!R^2 \to I\!R^2$ is a map such that $R \circ R = 1$ and the Jacobian determinant of R is everywhere negative. Most area-preserving maps of the plane have such a symmetry, including the area-preserving Hénon map of the previous section and Hamiltonian maps to be discussed in a subsequent chapter.

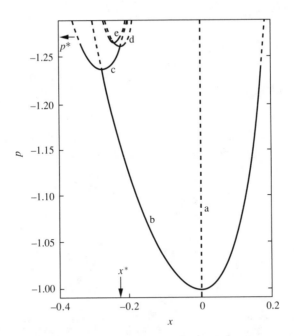

FIGURE 7.18   A section of the period-doubling tree of the De Vogelaere quadratic map in the plane $y=0$ showing the fixed point with its branches. A continuous line indicates stability and a dashed line indicates instability.

As two simple, but useful examples $(x, y) \mapsto (x', y')$ consider

$$S_1: \mathbb{R}^2 \to \mathbb{R}^2, \qquad (x, y) \mapsto (x, -y). \qquad (7.54)$$

$$S_2: \mathbb{R}^2 \to \mathbb{R}^2, \qquad (x, y) \mapsto (y + f(x), x - f(x')). \qquad (7.55)$$

Note the implicit nature of the map in the y-slot in (7.55). We remark that the map given by

$$F = S_1 \circ S_2 \qquad (7.56)$$

has a symmetry since both S_1 and S_2 are reflections. Furthermore, Exercise 7.7 shows that F of (7.56) is equivalent to the area-preserving Hénon map for an appropriate choice of $f(x)$. Thus F is equivalent to the map

$$(x, y) \mapsto (-y + f(x), x - f(x')). \qquad (7.57)$$

Greene et al. (1981) have chosen

$$f(x) = px - (1 - p)x^2 \qquad (7.58)$$

and have referred to this map as the De Vogelaere quadratic map. This map is equivalent to the area-preserving Hénon map. The advantage of studying the area-preserving Hénon map in this form is that the dominant symmetry curve is straighened out and the coordinate scalings are more apparant.

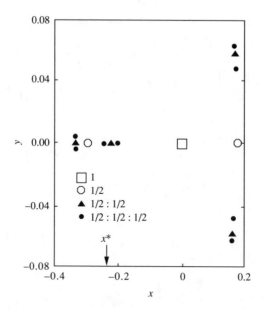

FIGURE 7.19   Section of the period-doubling tree of the De Vogelaere quadratic map for $p=p^*$. The positions of the points in the $x-y$ plane after each successive bifurcation are indicated by the symbols in the legend.

Figure 7.18, taken from Greene et al. (1981), shows the bifurcation tree of the De Vogelaere quadratic map as the parameter p is increased in the $y = 0$ plane. Figure 7.19 taken from the same reference, shows the first few points in the bifurcation sequence. The parameter values for which the bifurcations take place accumulate at the value $p^* = -1.26631127692\ldots$ (c_∞ of the previous section), and the point on the $y = 0$ line to which the 2^n-cycle points accumulate is given by $x^* = -0.23600609\ldots$. The sequence of the p_n converges geometrically to the rescaling factor δ given by

$$\lim_{n\to\infty} \frac{p_n - p_{n+1}}{p_{n+1} - p_{n+2}} = \delta = 8.721097200\ldots\;.$$

Refer to Fig. 7.18 and let $(x_n(0), 0)$ be the point of the 2^n-cycle on the $y = 0$ line that doubles along the $y = 0$ line at p_n. We denote the other point along the $y = 0$ line as $\left(x_n(\frac{1}{2}), 0\right)$, which is halfway around the orbit from $(x_n(0), 0)$. Then

$$\lim_{n\to\infty} \frac{x_n(0) - x_n(\frac{1}{2})}{x_{n+1}(0) - x_{n+1}(\frac{1}{2})} = \alpha = -4.018076704\ldots\;.$$

We can also look for rescaling across the symmetry line $y = 0$. Now fix $p = p^*$ and refer to Fig. 7.19. Let $(x_n^*(0), 0)$ denote the position of the point that doubled along the line $y = 0$ at $p = p^*$. Then $(x_n^*(\frac{1}{2}), 0)$ denotes the points halfway around as before and $\left(x_n^*(\frac{1}{4}), y_n^*(\frac{1}{4})\right)$ and $\left(x_n^*(\frac{3}{4}), y_n^*(\frac{3}{4})\right)$ denote the points $\frac{1}{4}$ and $\frac{3}{4}$ of the way around the orbit. These points are symmetric with respect to the $y = 0$ symmetry line. Then again

$$\lim_{n\to\infty} \frac{x_n^*(0) - x_n^*(\frac{1}{2})}{x_{n+1}^*(0) - x_{n+1}^*(\frac{1}{2})} = \alpha,$$

and across the line we have a new rescaling factor

$$\lim_{n\to\infty} \frac{y_n^*(\frac{1}{4})}{y_{n+1}^*(\frac{1}{4})} = \beta = 16.363896879\ldots\ .$$

The whole pattern of periodic points of period 2^n repeats itself asymptotically on rescaling by α and β.

Thus, we now have an operator on maps that is analogous to the doubling operator of Chapter 3. Following the notation of Chapter 3, we denote this operator as T and write

$$T(F) = B \circ F^2 \circ B^{-1}, \qquad B = \begin{pmatrix} \alpha & 0 \\ 0 & \beta \end{pmatrix}. \tag{7.59}$$

The remarkable feature is that the parameters α, and β are universal, and there is a map F^* that is an attracting fixed point of the doubling operator T in (7.59). Greene et al. (1981) give the following approximation for the universal attractor F^*:

$$x' = 0.405 - 0.1947x - 2.2676x^2 - 1.006y + \cdots$$

$$y' = 0.0502 + 0.5964x - 4.59x^2 - 2.056y + \cdots.$$

Certainly the last word on universality in maps has not been said. Higher dimensional maps undergoing period-doubling have also been shown to have universal properties (Mao and Greene, 1987). Also universal properties of dissipative maps have been studied by Chen et al. (1987). It is certain that universality is to remain an active and fruitful area of research for some time.

7.7 Summary

In this chapter we have seen that two-dimensional maps have bifurcation structures even richer than those for one-dimensional maps. A new feature in two-dimensional maps is Hopf-bifurcation and the occurrence of resonances. Arnold tongues were encountered describing regions in parameter space where the resonances can persist. Where these tongues overlap chaos sets in. The Hénon map is probably the most well-studied two-dimensional map and led, in its area-preserving form, to an introduction to universality for area-preserving maps.

Exercises

7.1 Verify that in the normal form the term $w_n^2 w_n^*$ is always in resonance and that the $1{:}3$ resonance leads to a term w_n^{*2} and the $1{:}4$ resonance leads to a term w_n^{*3}.

7.2 For $a > 0$, $B\cos(\beta - \alpha)$, and $\mu > 0$ in (7.28), show that R_- is stable and R_+ is unstable for the crater–Hopf bifurcation.

7.3 Derive Eq. (7.32).

7.4 Show that the rectangular region defined by the points $(-1.33, 0.42)$, $(1.32, 0.133)$, $(1.245, -0.14)$, $(-1.06, -0.5)$ is mapped inside itself under the Hénon map. This rectangular region thus serves as a trapping region for the Hénon map.

7.5 Show that the fixed points of (7.43) are given by $x = (-1 \pm \sqrt{1+a})/a$. Transform the origin to the negative fixed point and show that (7.43) can be written in the form (7.44).

7.6 Show that for the Hénon map (7.40) only the area-preserving case $b = -1$ can be renormalized in the manner described.

7.7 Show that S_1 and S_2 of (7.54) and (7.55) are reflections. Show that F of (7.56) is equivalent to the area-preserving Hénon map of (7.44). [Hint: Use the area-preserving coordinate change $X = x$, $Y = y + f(x)$ and choose an appropriate $f(x)$.]

CHAPTER EIGHT

CONSERVATIVE DYNAMICS

*... there is a God precisely because Nature
itself, even in chaos, cannot proceed except in
an orderly and regular manner.*

Immanuel Kant

At the conclusion of the last chapter our attention had shifted to a special subclass of two-dimensional maps, those that preserved area. As it turns out, the
familiar Liouville theorem for Hamiltonian mechanics ensures that mappings
associated with Hamiltonian systems preserve phase-space volume. For two-
dimensional examples this means preservation of area. As a consequence the
entire issue of attracting sets becomes irrelevant for Hamiltonian dynamics.
There is no such thing as a strange attractor for a Hamiltonian system.

Despite this seemingly severe restriction to a rather small subclass of dynamical systems, namely the nondissipative ones, there are numerous applications and examples of chaotic dynamics in conservative dynamical systems.
The motion of charged particles in electromagnetic fields, as in accelerators or
plasma experiments, is one important example. The geometries of magnetic
field lines is another. The voluminous literature dealing with this topic is a
further testimony to the importance of this aspect of chaotic dynamics.

In this single chapter we cannot hope to explore thoroughly all important
aspects of this area but rather aim to introduce the reader to some of the
results, issues, and methods. For the reader who desires more detail, we
recommend the monograph by Lichtenberg and Lieberman (1983) and the
review article by Helleman (1980). Both of these references contain extensive
references to the research literature.

In our introduction to the subject of chaos in conservative dynamics, we
review the basic concepts of Hamiltonian dynamics, primarily to establish the
notation. For a detailed development of Hamiltonian dynamics we refer the
reader to a textbook such as Rasband (1983), Arnold (1978), or Goldstein
(1980). The notation used in this chapter corresponds to the first of these
references.

After establishing the notation with a short introduction to Hamiltonian
dynamics and canonical variables for integrable systems, nearly integrable
Hamiltonian systems are considered. We review the implications of the KAM
theorem and the effect of resonances. Standard examples are used to illus-

161

trate these concepts, and the chapter concludes with a discussion of Arnold diffusion.

8.1 Hamiltonian Dynamics and Transformation Theory

For reasons of brevity, and in the expectation that the reader has previously encountered a thorough discussion of Lagrangian and Hamiltonian dynamics, we present a review only of the major results.

The Hamiltonian formulation is a powerful and effective technique for analyzing dynamical problems. The Hamiltonian function is usually defined in terms of the Lagrangian function:

$$H(\mathbf{p}, \mathbf{q}, t) = p_i \dot{q}^i - L(\mathbf{q}, \dot{\mathbf{q}}, t), \tag{8.1}$$

where the repeated index i denotes a sum over $i = 1, \ldots, n$ as usual. The coordinates $\mathbf{q} = (q^1, q^2, \ldots, q^n)$ represent the *generalized coordinates* of the system under study (with n degrees of freedom). The quantities $\dot{\mathbf{q}} = (\dot{q}^1, \dot{q}^2, \ldots, \dot{q}^n)$ are the *generalized velocities* with $\dot{q}^i \equiv dq^i/dt$. The *canonical momenta* p_i are obtained from the Lagrangian according to

$$p_i = \frac{\partial L}{\partial \dot{q}^i}. \tag{8.2}$$

The Lagrangian is obtained as the difference between the kinetic and potential energies in the usual way.

$$L(\mathbf{q}, \dot{\mathbf{q}}, t) = T(\dot{\mathbf{q}}) - U(\mathbf{q}, t). \tag{8.3}$$

We use (8.2) to find $\dot{q}^i = \dot{q}^i(\mathbf{p}, \mathbf{q})$ and thereby eliminate the dependence in (8.1) on velocities in favor of a dependence on canonical momenta and generalized coordinates. Hamilton's dynamical equations for the motion are

$$\dot{q}^i = \frac{\partial H}{\partial p_i}, \qquad \dot{p}_i = -\frac{\partial H}{\partial q^i}, \quad i = 1, \ldots, n, \tag{8.4}$$

with $\partial H/\partial t = -\partial L/\partial t$.

The canonical momenta $\mathbf{p}$ and the coordinate $\mathbf{q}$ serve as coordinates in a phase space for the system. Modern expositions of dynamics identify this phase space with the *cotangent bundle* T^*M, where M denotes the configuration manifold for the dynamical system under study. Hamilton's equations (8.4) show that the dynamical evolution of the system generates a trajectory in phase space for which the components of the tangent vector are given by partial derivatives of the Hamiltonian $H(\mathbf{p}, \mathbf{q})$. That is,

$$\mathbf{v}_H = \dot{p}_i \mathbf{e}_{p_i} + \dot{q}^i \mathbf{e}_{q^i} = -\frac{\partial H}{\partial q^i} \mathbf{e}_{p_i} + \frac{\partial H}{\partial p_i} \mathbf{e}_{q^i} \tag{8.5}$$

is a tangent vector along the trajectories in phase space T^*M, and the vectors $\{\mathbf{e}_{p_i}, \mathbf{e}_{q^i}\}$ form the usual set of coordinate and momentum basis vectors. The vector $\mathbf{v}_H$ is called the *flow-vector field* for the function $H(\mathbf{p}, \mathbf{q})$.

As with all manifolds, however, there is more than one way to put coordinates on T^*M. A transformation of coordinates $(\mathbf{p}, \mathbf{q}) \rightarrow (\mathbf{P}, \mathbf{Q})$ such that there is a scalar function $H'(\mathbf{P}, \mathbf{Q})$ determining the $\dot{P}_i$ and $\dot{Q}^i$ as in (8.4) is called a *canonical transformation*.

A convenient and compact way of characterizing canonical transformations makes use of the differential 2-form defined on phase space T^*M.

$$\boldsymbol{\gamma} = \mathbf{d}p_i \wedge \mathbf{d}q^i, \tag{8.6}$$

where the $\mathbf{d}p_i$ and $\mathbf{d}q^i$ are the usual differential 1-forms that constitute a coordinate basis for the covectors on the manifold M and $\wedge$ denotes the exterior (wedge) product. [No calculations using the exterior product or the exterior derivative $\mathbf{d}$ will be pursued in this chapter and so the reader for whom these concepts are unfamiliar may simply view them as oddities of notation and refer to familiar definitions for canonical transformations and generating functions. A readable source for a discussion of exterior algebra with differential forms is Schutz (1980).] A time-independent transformation $(p_i, q^i) \rightarrow (P_i, Q^i)$ is canonical if and only if

$$\boldsymbol{\gamma} = \mathbf{d}p_i \wedge \mathbf{d}q^i = \mathbf{d}P_i \wedge \mathbf{d}Q^i. \tag{8.7}$$

An equivalent statement in terms of a generating function S_1 is that

$$p_i \mathbf{d}q^i = P_i \mathbf{d}Q^i + \mathbf{d}S_1 \tag{8.8}$$

since the exterior derivative $\mathbf{d}$ applied to (8.8) gives

$$\boldsymbol{\gamma} = \mathbf{d}(p_i \mathbf{d}q^i) = \mathbf{d}p_i \wedge \mathbf{d}q^i = \mathbf{d}(P_i Q^i) + \mathbf{d}^2 S_1 = \mathbf{d}P_i \wedge \mathbf{d}Q^i,$$

where we recall that $\mathbf{d}^2 = 0$ for exterior differentiation. Using $S_1(\mathbf{q}, \mathbf{Q})$ one then obtains from (8.8)

$$\frac{\partial S_1}{\partial q^i} = p_i, \qquad \text{and} \qquad \frac{\partial S_1}{\partial Q^i} = -P_i. \tag{8.9}$$

Other types of generating function relations can be obtained in a similar way after performing a Legendre transformation on (8.8). For further detail and a discussion of the time dependent case we refer the reader to Chapter 8 of Rasband (1983).

An important consequence of Hamiltonian flow on phase space T^*M is that the flow is incompressible and hence preserves volumes. This is known as Liouville's theorem and characterizes an important difference between Hamiltonian dynamics and dissipative dynamics.

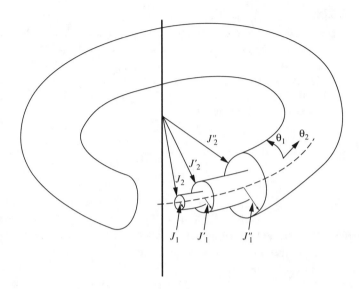

FIGURE 8.1　　A representation of an invariant torus for a system with two degrees of freedom.

As special as the class of Hamiltonian systems are in the set of all dynamical systems, there is an even more restricted class that gets most of the attention in first courses on this subject. This class is that of *integrable* Hamiltonian systems. An integrable Hamiltonian system with n degrees of freedom has n first integrals of the motion $I_i, i = 1, \ldots, n$. They must be linearly independent, single-valued functions on phase space. The flow vectors they generate $\mathbf{v}_{I_i}, i = 1, \ldots, n$, must everywhere commute. If in addition the motion is bounded, then it may be represented as a trajectory on the surface of an n-dimensional torus. This n-dimensional torus, determined by a given choice of values for the first integrals I_i, is then the submanifold of T^*M on which the motion takes place. On such a torus exactly n closed paths can be chosen that cannot be deformed into one another or to a point. See Fig. 8.1 for an illustration of the T^2 torus. On the surface of such a T^2 torus one path goes the long way around (toroidal) and the other path goes the short way around (poloidal).

If we denote such a set of paths by $\{\Gamma_i\}$, then the *action momenta* are defined by

$$J_i = \frac{1}{2\pi} \oint_{\Gamma_i} p_j(\mathbf{I}, \mathbf{q}) \, dq^j, \quad i = 1, \ldots, n. \tag{8.10}$$

Note the sum on j. The integal invariants $\{I_i\}$ can now all be expressed in terms of the action momenta, which form an equivalent set of first integrals for the motion. The function

$$S_2(\mathbf{q}, \mathbf{J}) = \int_{\mathbf{q}_0}^{\mathbf{q}} p_j(\mathbf{J}, \mathbf{q}') \, dq'^j \tag{8.11}$$

serves as a type 2 generating function for a canonical transformation to the coordinates $(\mathbf{J}, \theta)$. The coordinates

$$\theta^i = \frac{\partial S}{\partial J_i} \qquad (8.12)$$

fill the role of angle coordinates on the n-tori in that they increase by 2π in going once around the closed curve Γ_i.

For such a completely integrable system the Hamiltonian is only a function of the action momenta, $H = H(\mathbf{J})$. As a consequence, Hamilton's equations of motion become

$$\dot{J}_i = -\frac{\partial H}{\partial \theta^i} = 0 \qquad (8.13)$$

$$\dot{\theta}^i = \frac{\partial H}{\partial J_i} = \omega^i(\mathbf{J}) = \text{constant}, \qquad (8.14)$$

with the obvious solution

$$J_i = \text{constant}, \qquad \theta^i(t) = \omega^i(\mathbf{J})t + \theta_0^i. \qquad (8.15)$$

8.2 Two Examples in Action-Angle Coordinates

As an illustration of these ideas we cite two examples, the details of which can be found in the dynamics texts previously referenced. The first example is the simple harmonic oscillator. This example is a small amplitude approximation to the simple pendulum, which is the second example.

The Hamiltonian of the simple harmonic oscillator is taken in the form

$$H(p, q) = \frac{p^2}{2} + \frac{\omega_0^2 q^2}{2}. \qquad (8.16)$$

For this case phase space T^*M with coordinates (p, q) can be identified with the manifold $\mathbb{R}^2$, the familiar plane. The trajectories in phase space consist of ellipses, each ellipse labeled by a constant value of H. See Fig. 8.2. For this one-degree of freedom system H is the only first integral. For this system it is clear that the n-tori ($n = 1$) are just the ellipses and that each ellipse is labeled by a different value for H. From (8.16), $p = \pm\sqrt{2H - \omega_0^2 q^2}$, and from (8.10)

$$J = \frac{1}{2\pi} \oint p\, dq = \frac{\text{area of ellipse}}{2\pi} = \frac{H}{\omega_0}. \qquad (8.17)$$

The action angle is obtained from (8.11) with (8.12),

$$\theta = \frac{\partial}{\partial J} \int_0^q (2J\omega_0 - \omega_0^2 q'^2)^{\frac{1}{2}}\, dq'.$$

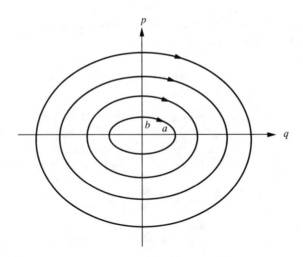

FIGURE 8.2   Elliptical phase curves for the one-dimensional harmonic oscillator of (8.16). The arrows show the direction of traversal for the trajectories. The semi-axes for the ellipse are given by $a=\sqrt{2H}/\omega_0=\sqrt{2J/\omega_0}$ and $b=\sqrt{2H}=\sqrt{2J\omega_0}$.

There results

$$q = \sqrt{\frac{2J}{\omega_0}} \cos\theta, \qquad p = -\sqrt{2J\omega_0}\sin\theta, \tag{8.18}$$

and we see directly that

$$H = J\omega_0 \quad \text{and} \quad \dot\theta = \frac{\partial H}{\partial J} = \omega_0, \tag{8.19}$$

are consistent with (8.13) and (8.14). In this case the angular frequency $\omega(J) = \omega_0$ is constant, i.e., it is the same on every invariant curve. From Hamilton's equations (8.4) the familiar linear differential equation $\ddot{q}+\omega_0^2 q = 0$ is readily obtained.

In contrast to the previous example, the simple pendulum is not linear, and it follows that the angular frequency is not a constant but rather is dependent on the action momentum J (or the energy H). The equation of motion for the simple pendulum may be written in the form

$$\ddot\phi + \omega_0^2 \sin\phi = 0, \tag{8.20}$$

where ϕ represents the angular displacement in a counterclockwise direction from the lower equilibrium position. Letting $p = \dot\phi = \partial H/\partial p$ and $\dot{p} = -\omega_0^2 \sin\phi = -\partial H/\partial\phi$ results in the Hamiltonian

$$H = \frac{p^2}{2} - \omega_0^2 \cos\phi. \tag{8.21}$$

The range of the coordinate ϕ is from $-\pi$ to π. Figure 8.3 shows a sketch of the phase space manifold.

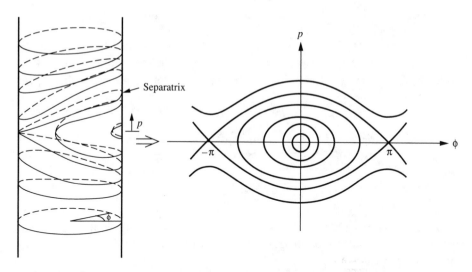

FIGURE 8.3   The sketch on the left shows the phase-space manifold for the simple pendulum. Often this manifold is cut along the $\phi = \pm\pi$ line and unrolled to give the sketch on the right.

A complete solution of the simple pendulum requires the use of elliptic functions. For formulae and results involving elliptic functions we recommend as references Byrd and Friedman (1954) or Abramowitz and Stegun (1972). It is straightforward to verify that a solution to (8.20) is given by

$$\sin\frac{\phi}{2} = k\,\mathrm{sn}(\omega_0 t, k), \qquad (8.22)$$

where sn, cn, and dn are the Jacobi elliptic functions in the usual notation. Using the result that $\cos\phi/2 = \mathrm{dn}(\omega_0 t, k)$ and $\partial\,\mathrm{sn}(\psi, k)/\partial\psi = \mathrm{cn}(\psi, k)\,\mathrm{dn}(\psi, k)$, we find

$$p = \dot{\phi} = 2k\omega_0\,\mathrm{cn}(\omega_0 t, k). \qquad (8.23)$$

Substitution of the foregoing results into (8.21) shows that

$$k^2 = \frac{1}{2}\left(\frac{H}{\omega_0^2} + 1\right) \qquad (8.24)$$

and k goes from 0 to 1 as H goes from $-\omega_0^2$ to $+\omega_0^2$. These are the values for bounded motion. For the constant $H > \omega_0^2$ one obtains free rotation and no pendulum action. These orbits are shown clearly in Fig. 8.3 as those encircling fully the cylindrical surface representing the phase space.

We compute the action momentum in the usual way for a bounded (in ϕ) trajectory.

$$J = \frac{1}{2\pi}\oint p\,d\phi = \frac{2}{\pi}\int_0^{\phi_0}(2H + 2\omega_0^2\cos\phi)^{\frac{1}{2}}\,d\phi = \frac{4k\omega_0}{\pi}\int_0^{\phi_0}\mathrm{cn}(\omega_0 t, k)\,d\phi,$$

$$(8.25)$$

where ϕ_0 is the maximum value obtained by ϕ for a given $H < \omega_0^2$.

This result (8.25) can be written in terms of elliptic integrals. The maximum angle ϕ_0 is given in terms of the quarter period t_0 according to $\text{sn}(\omega_0 t_0, k) = \frac{1}{k}\sin\frac{\phi_0}{2}$. From (8.21) and (8.24) we have $\frac{1}{k}\sin\frac{\phi_0}{2} = 1$ and so $\arcsin(\frac{1}{k}\sin\frac{\phi_0}{2}) = \frac{\pi}{2}$. Resorting to the definition of sn in terms of elliptic integrals, we have

$$\omega_0 t_0 = \int_0^{\frac{\pi}{2}} (1 - k^2 \sin^2 \theta)^{-\frac{1}{2}} d\theta = F\left(\frac{\pi}{2}, k\right) = K, \qquad (8.26)$$

where

$$F(\psi, k) = \int_0^{\psi} (1 - k^2 \sin^2 \theta)^{-\frac{1}{2}} d\theta$$

is the elliptic integral of the first kind and $F(\frac{\pi}{2}, k) \equiv K$ is the complete elliptic integral of the first kind. Thus we have that the period of the motion is given according to

$$T(k) = 4K/\omega_0, \qquad (8.27)$$

and the frequency by $\omega(k) = 2\pi/T = \omega_0 \pi / 2K$, where K depends on the modulus k, which in turn depends on the energy constant H.

Returning to the action in (8.25), we change the variable of integration to $\omega_0 t$. From (8.22)

$$d\phi = 2k\,\text{cn}(\omega_0 t, k)d(\omega_0 t)$$

$$J = \frac{8k^2\omega_0}{\pi} \int_0^K \text{cn}^2(\psi, k)\, d\psi = \frac{8\omega_0}{\pi}(E - (1 - k^2)K), \qquad (8.28)$$

where E is the complete elliptic integral of the second kind. Clearly J is only a function of k, which is in turn a function of H, and vice versa. The frequency

$$\omega(J) = \frac{dH}{dJ} = \frac{dH}{dk}\left(\frac{dJ}{dk}\right)^{-1},$$

where

$$\frac{dJ}{dk} = \frac{8\omega_0}{\pi}\left(\frac{dE}{dk} - (1 - k^2)\frac{dK}{dk} + 2kK\right).$$

Using the results $dE/dk = (E - K)/k$ and $dK/dk = (E - (1 - k^2)K)/k(1 - k^2)$, it is straightforward to show that (8.27) results. In this case $\omega(J)$ and J depend parametrically on k.

8.3 Nonintegrable Hamiltonians

The two foregoing sections outline the theory of integrable Hamiltonians and the description of the motion generated by such Hamiltonians. As pretty and satisfying as this theory is, it usually does not apply. Most Hamiltonians with more than one degree of freedom are not completely integrable. Generally, the only integral of the motion is the energy. For such a general Hamiltonian numerical integration is usually the only recourse for understanding the system, unless the general Hamiltonian is a small perturbation of a completely integrable Hamiltonian.

To illustrate these remarks and to contrast the integrable with the nonintegrable case, we consider two Hamiltonians that are related. The first Hamiltonian is referred to as the Hamiltonian for the Toda lattice (Ford, 1975).

$$H_T = \frac{1}{2}(p_x^2 + p_y^2) + \frac{1}{24}(e^{2x-2y\sqrt{3}} + e^{2x+2y\sqrt{3}} + e^{-4x}) - \frac{1}{8}. \qquad (8.29)$$

Since H_T is not an explicit function of time, H_T is a constant of the motion. Hamiltonian's equations of motion automatically give a differential system in first-order form:

$$\dot{x} = \frac{\partial H_T}{\partial p_x} = p_x, \qquad \dot{y} = \frac{\partial H_T}{\partial p_y} = p_y,$$

$$\dot{p}_x = -\frac{\partial H_T}{\partial x} = \frac{1}{6}[e^{-4x} - e^{2x}\cosh(2y\sqrt{3})], \qquad (8.30)$$

$$\dot{p}_y = -\frac{\partial H_T}{\partial y} = -\frac{e^{2x}}{2\sqrt{3}}\sinh(2y\sqrt{3}).$$

With H_T constant a system trajectory can be expected to fill out a three-dimensional volume forming the energy hypersurface in the four-dimensional phase space. We obtain a cross section of this volume by looking at a Poincaré section. We choose the section plane to be the (x, p_x) plane. With H_T fixed, every time the trajectory passes through the $y = 0$ plane with $p_y > 0$, we place a point at the corresponding (x, p_x) coordinate. Figure 8.4 shows the plots obtained for two different values of the energy H_T. Each closed curve in Fig. 8.4 corresponds to a different initial point for the trajectory.

The striking thing about the curves in Fig. 8.4 is that the trajectories clearly lie on surfaces. The curves do not fill out a three-dimensional volume as initially expected. This suggests that there is another integral of the motion in addition to the energy H_T. This second integral of the motion was found by Hénon (1969).

$$I = 8\dot{y}(\dot{y}^2 - 3\dot{x}^2) + (\dot{y} + \dot{x}\sqrt{3})e^{2x-2y\sqrt{3}} + (\dot{y} - \dot{x}\sqrt{3})e^{2x+2y\sqrt{3}} - 2\dot{y}e^{-4x}. \quad (8.31)$$

This integral is hard to find, but once given, it is straightforward to check that it is indeed constant.

Thus, knowing that H_T of (8.29) and I of (8.31) are both independent constants of the motion for (8.30), the curves of Fig. 8.4 are easily understood; they simply represent the intersection of nested tori like those in Fig. 8.1 with the (x, p_x) plane.

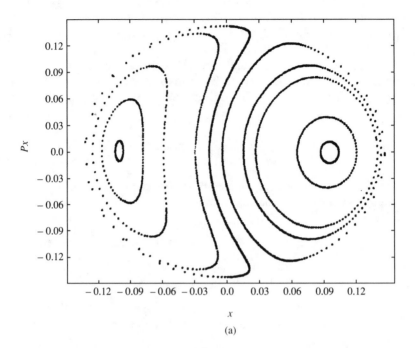

(a)

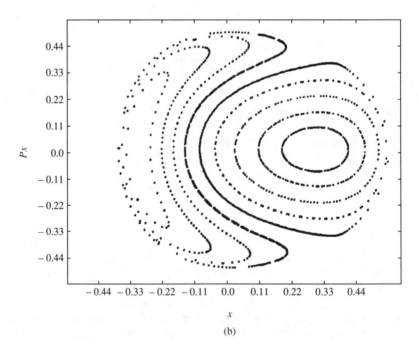

(b)

FIGURE 8.4   Poincare section plots for the system of (8.29): (a) $H_T = 0.01$ and (b) $H_T = 0.12$. Each of the nested curves is determined by the intersections of a phase-space trajectory with the (x, p_x) plane. Different curves correspond to different initial values.

For the second Hamiltonian system considered in this section, we expand (8.29) in the coordinates (x, y) about (0,0). As is easily seen, the lowest order terms correspond to the potential for a simple harmonic oscillator in two dimensions. The next order term corresponds to a cubic potential, and we drop all terms beyond this one. There results the Hénon and Heiles (1964) Hamiltonian:

$$H_e = \frac{1}{2}(p_x^2 + p_y^2) + \frac{1}{2}(x^2 + y^2) + y^2 x - \frac{1}{3}x^3. \tag{8.32}$$

In comparison to H_T in (8.29), this Hamiltonian seems much simpler. However, the polynomial nature of the potential does not translate into simpler dynamics. Figure 8.5 gives Poincaré section plots for the Hamiltonian of (8.32) for exactly the same energies chosen for the Toda Hamiltonian H_T in the section plots of Fig. 8.4.

The striking thing about the curves in Fig. 8.5 is the complexity of the island structures and the clearly chaotic orbit. The points in the chaotic region of Fig. 8.5 were all made by the intersections of a single trajectory.

There are several lessons to be learned from these two examples: (1) "Simple" Hamiltonians can result in complex dynamics. (2) The trajectories in phase space can indeed fill out a significant fraction of the energy hypersurface (in this case a 3-volume) if not constrained by additional integrals of the motion. (3) Small changes to completely integrable Hamiltonians can make significant qualitative changes in the dynamics.

The islands in Fig. 8.5 are produced by resonances. In the following section we see how resonant perturbations lead to island structures around the invariant tori and find that as the perturbations increase in strength, the invariant tori can be completely destroyed.

8.4 Canonical Perturbation Theory and Resonances

Most often the analysis of nonintegrable Hamiltonians proceeds via perturbation of integrable Hamiltonians. For this purpose we consider the perturbation theory of Hamiltonians in canonical variables. There are other methods of dealing with perturbations, such as Lie transforms, that are discussed in the references cited. In a particular situation one method may be preferable over another, but canonical perturbation theory illustrates clearly the issues associated with resonances, and we confine our attention to this method.

We consider a Hamiltonian partitioned into a piece that is completely integrable and a piece considered to be a perturbation.

$$H(\mathbf{J}, \theta) = H_0(\mathbf{J}) + \epsilon H_1(\mathbf{J}, \theta). \tag{8.33}$$

The action-angle variables $(\mathbf{J}, \theta)$ are the result of solving the dynamics problem with the Hamiltonian H_0. We consider the split represented by (8.33) to

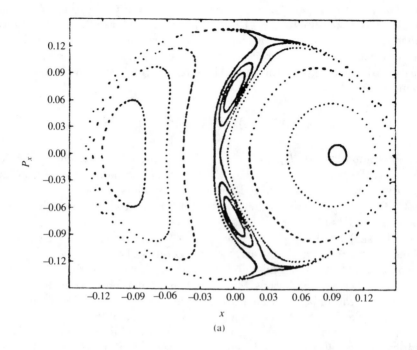

(a)

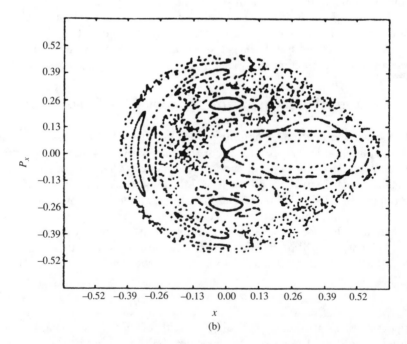

(b)

FIGURE 8.5   Poincaré section plots for trajectories of the Hénon–Heiles Hamiltonian (8.32): (a) $H_e = 0.01$ and (b) $H_e = 0.12$.

be valid when the system point in phase space is near the invariant torus of H_0 with the label $\mathbf{J}$. The phase space for H_0 is the same as the phase space for H. The variables $(\mathbf{J}, \theta)$ are a set of canonical variables for H; they are not, however, a set of action-angle variables, except for H_0, since the full Hamiltonian H depends on θ as well as on $\mathbf{J}$. The quantity ϵ formally measures the strength of the perturbation, but it may be nothing more than an artificial label to be set equal to one at the conclusion of the calculation.

If the perturbation is small, we might expect the existence of a set of canonical variables $(\mathbf{I}, \psi)$ for which H is completely integrable. One way to search for $(\mathbf{I}, \psi)$ is to try to find the canonical transformation relating $(\mathbf{J}, \theta)$ and $(\mathbf{I}, \psi)$. To this end we consider an expansion of the generating function for this desired transformation. Again we consider a type 2 generating function of new momenta and old coordinates.

$$S(\mathbf{I}, \theta) = \theta^i I_i + \epsilon S_1(\mathbf{I}, \theta) + \cdots . \tag{8.34}$$

For a type 2 generating function the new coordinates and the old momenta are given by

$$\psi^i = \frac{\partial S}{\partial I_i} = \theta^i + \epsilon \frac{\partial S_1}{\partial I_i} + \cdots \qquad \text{and} \qquad J_i = \frac{\partial S}{\partial \theta^i} = I_i + \epsilon \frac{\partial S_1}{\partial \theta^i} + \cdots . \tag{8.35}$$

The perturbation Hamiltonian $H_1(\mathbf{J}, \theta))$ is known and must be a periodic function of the angles $\theta = (\theta^1, \theta^2, \ldots, \theta^n)$. Consequently, we expand H_1 in a Fourier series:

$$H_1(\mathbf{J}, \theta) = \sum_{\mathbf{m}} H_{1\mathbf{m}}(\mathbf{J}) e^{i \mathbf{m} \cdot \theta}, \tag{8.36}$$

where $\mathbf{m} = (m_1, m_2, \ldots, m_n)$ and the m_i are integers. The perturbation function $S_1(\mathbf{I}, \theta)$ can also be expanded:

$$S_1(\mathbf{I}, \theta) = \sum_{\mathbf{m} \neq 0} S_{1\mathbf{m}}(\mathbf{I}) e^{i \mathbf{m} \cdot \theta}. \tag{8.37}$$

Exclusion of the constant term in (8.37) is to make it consistent with (8.34), so that the perturbation to the generating function only gives corrections to the variables in (8.35).

To express $H_0(\mathbf{J})$ in terms of $\mathbf{I}$ and ψ, we make a Taylor expansion in ϵ:

$$H_0(\mathbf{J}) = H_0(\mathbf{I}) + \epsilon \frac{\partial H_0}{\partial J_i} \bigg|_{\mathbf{J}=\mathbf{I}} \frac{\partial S_1}{\partial \theta^i} + \cdots , \tag{8.38}$$

where we have used (8.35). To lowest order in ϵ

$$\frac{\partial H_0}{\partial J_j} \bigg|_{\mathbf{I}} = \frac{\partial H_0}{\partial J_j} \bigg|_{\mathbf{J}} = \omega^j(\mathbf{J}) = \dot{\theta}^j . \tag{8.39}$$

Thus we have

$$H_0(\mathbf{J}) = H_0(\mathbf{I}) + \epsilon \omega^j(\mathbf{J})\frac{\partial S_1}{\partial \theta^j} + \cdots. \tag{8.40}$$

Using (8.37)

$$\frac{\partial S_1}{\partial \theta^j} = \sum_{\mathbf{m} \neq 0} i m_j S_{1\mathbf{m}}(\mathbf{I}) e^{i\mathbf{m}\cdot\theta}. \tag{8.41}$$

We substitute (8.41), (8.40), and (8.36) into (8.33) to obtain

$$H(\mathbf{I}) = H_0(\mathbf{I}) + \epsilon\left\{\sum_{\mathbf{m}\neq 0} i(\omega\cdot\mathbf{m})S_{1\mathbf{m}}(\mathbf{I})e^{i\mathbf{m}\cdot\theta} + \sum_{\mathbf{m}} H_{1\mathbf{m}}(\mathbf{I})e^{i\mathbf{m}\cdot\theta}\right\} + \cdots,$$

where to lowest order $H_{1\mathbf{m}}(\mathbf{J}) = H_{1\mathbf{m}}(\mathbf{I})$. Since by assumption $H(\mathbf{I})$ is not a function of θ, we equate to zero the coefficient of each $e^{i\mathbf{m}\cdot\theta}$ for each choice of $\mathbf{m}$. There results:

$$H(\mathbf{I}) = H_0(\mathbf{I}) + \epsilon H_{10}(\mathbf{I}) + \cdots$$
$$i(\omega(\mathbf{I})\cdot\mathbf{m})S_{1\mathbf{m}}(\mathbf{I}) + H_{1\mathbf{m}}(\mathbf{I}) = 0, \qquad \mathbf{m} \neq 0. \tag{8.42}$$

We solve the second of (8.42) to find

$$S_{1\mathbf{m}}(\mathbf{I}) = \frac{iH_{1\mathbf{m}}(\mathbf{I})}{\omega(\mathbf{I})\cdot\mathbf{m}}, \tag{8.43}$$

and

$$S_1(\mathbf{I},\theta) = \sum_{\mathbf{m}\neq 0} \frac{iH_{1\mathbf{m}}(\mathbf{I})}{\omega(\mathbf{I})\cdot\mathbf{m}} e^{i\mathbf{m}\cdot\theta}. \tag{8.44}$$

This expansion procedure is something we apply in the neighborhood of a particular n-torus of H_0 labeled with a particular value for $\mathbf{J}$. If for this particular $\mathbf{J}$ there is some $\mathbf{m}$ such that $(\omega\cdot\mathbf{m}) = 0$, then we have a resonance, and the corresponding invariant torus is referred to as a *resonant surface*.

Thus, there are actually two problems in expansion (8.34). There is the problem of convergence in ϵ, and there is also the problem of resonant denominators. For a sufficiently irrational $\omega(\mathbf{J})$ there is a convergent perturbation series in ϵ leading to a new action-angle set. This result is a conclusion of the celebrated KAM theorem, about which we have more to say later.

Let us now see what can be done about removing the difficulty of resonant denominators. For purposes of being explicit, we assume two degrees of freedom: $\theta = (\theta^1, \theta^2)$ and $\mathbf{J}^0 = (J_1^0, J_2^0)$. Furthermore, we assume that $r\omega^1(\mathbf{J}^0) - s\omega^2(\mathbf{J}^0) = 0$, where r and s have no common divisors other than 1. Thus the smallest $\mathbf{m}$ giving a resonant denominator is $\mathbf{m} = (r, -s)$.

We now choose new coordinates $\psi = (\psi^1, \psi^2)$ and $\mathbf{I} = (I_1, I_2)$ according to

$$\psi^1 = r\theta^1 - s\theta^2, \qquad \psi^2 = \theta^2, \tag{8.45}$$

$$I_1 = \frac{1}{r}(J_1 - J_1^0), \qquad I_2 = \frac{s}{r}(J_1 - J_1^0) + J_2 - J_2^0. \tag{8.46}$$

It is straightforward to check that this transformation is canonical by showing either $\mathbf{d}\theta^i \wedge \mathbf{d}J_i = \mathbf{d}\psi^i \wedge \mathbf{d}I_i$ or by showing that a generating function for this transformation is given by $S = \psi^i I_i + \theta^i J_i^0$.

This transformation makes the resonant surface at $\mathbf{J} = \mathbf{J}^0$ correspond to $\mathbf{I} = 0$. The rate of change of ψ^1, i.e., $\dot{\psi}^1 = r\dot{\theta}^1 - s\dot{\theta}^2$, measures the deviation from resonance, and near the resonant surface $\dot{\psi}^1 \ll \dot{\psi}^2 = \dot{\theta}^2$ Consequently, ψ^2 is a rapidly varying coordinate in comparison to ψ^1 and averaging with respect to the variable ψ^2 is appropriate. Thus $H_1(\mathbf{J}, \theta)$ in (8.36) becomes

$$\langle H_1 \rangle_{\psi^2} = \bar{H}_1(\mathbf{I}, \psi^1) = \sum_n \bar{H}_{1p}(\mathbf{I}) e^{in\psi^1}. \tag{8.47}$$

Since $\bar{H}_1$ is independent of ψ^2, I_2 is a constant. [I_2 of (8.46) is actually the first term in an expansion for an adiabatic invariant of the Hamiltonian $H(\mathbf{I}, \psi)$.] With I_2 constant the motion in the (I_1, ψ^1) coordinates is effectively motion with one degree of freedom. In fact $I_2 = 0$ and we may completely ignore it. In (8.47) we keep only $n = 0, \pm 1$ terms. It is generally the case that the terms diminish rapidly with increasing n.

As a consequence of this expansion near a resonance surface, and of the averaging over the fast variables, the Hamiltonian of (8.33) takes on the form

$$\bar{H}(\mathbf{I}, \psi) \simeq \bar{H}_0(\mathbf{I}) - \epsilon\bar{H}_1(\mathbf{I}) \cos \psi^1. \tag{8.48}$$

Expanding near the rational surface $\mathbf{I} = 0$ now gives

$$\bar{H}_0(\mathbf{I}) = \bar{H}_0(0) + I_i \left(\frac{\partial \bar{H}_0}{\partial I_i} \right)_0 + \frac{1}{2} I_i I_j \left(\frac{\partial^2 \bar{H}_0}{\partial I_i \partial I_j} \right)_0 + \cdots.$$

Discarding the constant term and using the fact that $I_2 = $ constant, we obtain

$$\bar{H}_0(\mathbf{I}) = I_1 \left(\frac{\partial \bar{H}_0}{\partial I_1} \right)_0 + \frac{1}{2} I_1^2 \left(\frac{\partial^2 \bar{H}_0}{\partial I_1^2} \right)_0 + \cdots. \tag{8.49}$$

Note that by the chain rule

$$\left(\frac{\partial \bar{H}_0}{\partial I_1} \right)_0 = \left(\frac{\partial H_0}{\partial J_i} \right)_0 \frac{\partial J_i}{\partial I_1} = r\omega^1(\mathbf{J}^0) - s\omega^2(\mathbf{J}^0) = 0. \tag{8.50}$$

Consequently, (8.48) becomes

$$\bar{H}(\mathbf{I}, \psi) \simeq \frac{1}{2} I_1^2 \left(\frac{\partial^2 \bar{H}_0}{\partial I_1^2} \right)_0 - \epsilon\bar{H}_1(0) \cos \psi^1. \tag{8.51}$$

By a trivial rescaling of the momentum variable, (8.51) can be written in the form (8.21) for a simple pendulum. Thus the dynamics of the simple pendulum describes the dynamics of *any* perturbed, integrable Hamiltonian system near a resonance. Thus it is no accident that the resonant island structures, seen for example in Fig. 8.5b, look like a series of repetitions of the phase curves of the simple pendulum in Fig. 8.3.

Now that we have seen how perturbations of completely integrable systems lead to pendulumlike dynamics in the neighborhood of the resonant surface, we shift to a more qualitative and geometrical description of the phase-space trajectories in the neighborhood of the resonance surface.

Again we confine our attention to an integrable system with two degrees of freedom. Referring to Fig. 8.1, we choose a surface of section to be the $\theta^2 = 0$ plane. If we specify J_2, then since the energy H is a constant, J_1 is also determined. Then each successive iteration of the section map advances the θ^1 coordinate by an angle $2\pi\alpha(J_1)$, where $\alpha(J_1)$ is some function of the action J_1. This map is similar to a circle map of Section 6.5 and is called the *twist map*. It can be written in the form

$$\begin{aligned} J_{n+1} &= J_n, \\ \theta_{n+1} &= \theta_n + 2\pi\alpha(J_{n+1}), \end{aligned} \tag{8.52}$$

where (J, θ) denote the coordinates (J_1, θ^1). If the integral surface J is such that $\alpha(J)$ is irrational, then any initial condition on the map leads to points under the mapping that densely cover the circle of radius J. If $\alpha(J)$ is rational, say $\alpha(J) = p/q$, where p and q are relatively prime, then every point on the circle of radius J is a periodic point of period q.

If we now add a perturbation to the Hamiltonian as in (8.33), we know this simple picture changes. What happens is the subject of the celebrated KAM theorem. However, neither the details of the theorem nor its proof are important for our purposes. The interested reader can find a more detailed discussion in the references cited. For present purposes it suffices to observe that the KAM theorem states that for sufficiently irrational surfaces and sufficiently weak perturbations, the surfaces are preserved under the perturbation and are not destroyed. This is, of course, not the case for the resonant, rational surfaces. For rational surfaces we follow an argument given in Lichtenberg and Lieberman (1983, Section 3.2b).

Figure 8.6 represents the sketch of a rational (p/q) surface and irrational KAM surfaces at both larger and smaller radii. We speak of "surfaces" but Fig. 8.6 can only depict the intersection of these surfaces with the Poincaré plane. These intersections we call KAM curves. For purposes of being specific in Fig. 8.6, we have assumed that the rational surface corresponds to a 1:3 resonance. We assume that $\alpha(J)$ is an increasing function of J so that points on the outer irrational surface rotate in a clockwise direction under q iterations of the twist map, and points on the inner irrational surface move in the counterclockwise direction. Under a perturbation the irrational surfaces are distorted (not shown in sketch), but still preserved. Between these KAM curves sketched in Fig. 8.6 must lie another curve, the points of which have

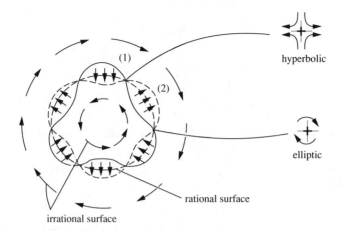

hyperbolic

elliptic

rational surface

irrational surface

FIGURE 8.6   Break-up of a rational surface under a perturbation into an even number of fixed points. See text for a discussion.

unaltered θ coordinate after q iterations of the perturbed twist map. We label this curve with a "1." Curve 1 is not a KAM curve. The points on curve 1 are mapped radially by the perturbed twist map to the points on curve "2.". Because the perturbed twist map must preserve area, curves 1 and 2 must enclose the same area. This is only possible if the curves 1 and 2 intersect in $2kq$ points (usually $k = 1$). These points are clearly fixed points of the perturbed twist map.

Figure 8.7 is a sketch of what the curves of intersection in the Poincaré plane might look like for the perturbed twist map. Around the elliptic fixed points are closed curves that can be thought of as the intersections with the Poincaré plane of a set of nested toroids wound around the original rational surface so as to intersect the Poincaré plane three times before closing.

Near the hyperbolic fixed points the dynamic trajectories are much more complicated. To understand qualitatively what happens, it helps to first consider the phase trajectories of the simple pendulum as sketched in the unrolled part of Fig. 8.3. There it is seen that part of the unstable manifold of the fixed point at $-\pi$ is the same as part of the stable manifold of the fixed point at π. The tendency is for the unstable and stable manifolds of the hyperbolic fixed points, as depicted in Fig. 8.7, to connect in the same way as for the pendulum. But this connection is structurally unstable and therefore the perturbation destroys it. The unstable manifold for the (usually) neighboring hyperbolic fixed point intersects transversally the stable manifold of the given hyperbolic fixed point an infinite number of times. This is sketched in the inset of Fig. 8.7. That the number of intersections is infinite follows because these intersection points asymptotically approach the given fixed point as the map is iterated an infinite number of times. Mapping behavior of this sort near an X point is clearly visible in one of the maps of Fig. 7.12.

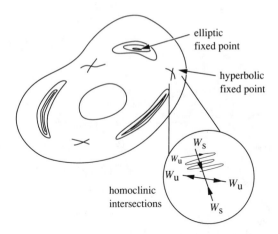

FIGURE 8.7   Qualitative sketch of the KAM surfaces of a perturbed integrable system.

The rapid oscillation as trajectories approach the hyperbolic points resulting from the perturbation of a rational surface is one way of viewing the emergence of chaotic regions around these hyperbolic points. These chaotic regions are evident in the Poincaré section of Fig. 8.5b as well as in Fig. 1.5.

8.5 The Standard Map

One of the most studied examples in the area of chaos in conservative systems is a map representative of a perturbed twist map:

$$J_{n+1} = J_n - \frac{k}{2\pi}\sin(2\pi\theta_n)$$
$$\theta_{n+1} = \theta_n + J_{n+1}. \tag{8.53}$$

This map is sometimes called the Taylor-Chirikov map and has been studied extensively. A major reason for the attention given to this map is that it occurs as an approximation to a large number of physical systems. For example consider again the differential equation for a simple pendulum in (8.20). Consider a naive numerical integration of this system and write centered-difference approximations to derivatives: $(\phi_{j+\frac{1}{2}} - \phi_{j-\frac{1}{2}})/\Delta t \simeq \dot{\phi}$. Then a difference equation approximation to (8.20) is simply

$$\phi_{j+1} - 2\phi_j + \phi_{j-1} = \frac{k}{2\pi}\sin(2\pi\phi), \tag{8.54}$$

where k has been suitably defined. We convert this second-order difference equation into a pair of first-order equations by defining $J_{j+1} \equiv \phi_{j+1} - \phi_j$. With this definition and (8.54) we obtain immediately the system (8.53). This relationship of the standard map to the simple pendulum and other dynamical

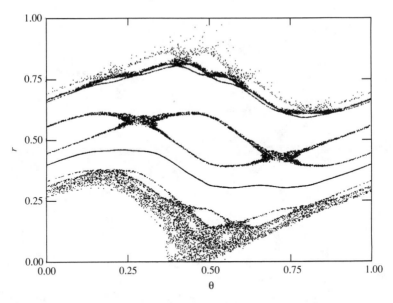

FIGURE 8.8   Orbits corresponding to five different initial conditions for the standard map of (8.53) with $k=0.97$.

systems has been discussed by Kadanoff (1985). In addition to Lichtenberg and Lieberman (1983) we refer the interested reader to Chirikov (1979) and Greene (1979) for detailed discussions of the standard map.

The mapping (8.53) is naturally periodic in the coordinates (J, θ) with period 1. The quantity k is to be considered the parameter measuring the strength of the perturbation. With $k = 0$, J is a constant of the motion, and all we get is perodic orbits if J is rational and an ergodic orbit it J is irrational.

For nonzero k we obtain similarly periodic orbits, ergodic orbits, and in addition chaotic orbits. Figure 8.8 shows typical orbits for a nonzero k. The periodic orbits are the elliptic points at the centers of the island chains. An ergodic orbit is manifest as a single distorted curve that crosses from one side to the other in Fig. 8.8. Chaotic orbits fill out a two-dimensional region in Fig. 8.8.

8.6 Arnold Diffusion

For dynamic systems certain behavior can only occur if the effective dimensionality of the system is large enough. For a conservative system with one degree of freedom, chaotic motion is not possible because the Hamiltonian is always a constant of the motion. In order to have chaotic motion it is necessary to have at least two degrees of freedom and hence to have at least a four-dimensional phase space. In four dimensions the Hamiltonian is again a

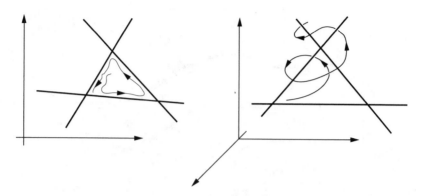

FIGURE 8.9   Phase-space trajectories are confined by lines (KAM surfaces) in two-dimensional space but not in three dimensions.

constant of the motion, and three dimensions remain for the chaotic trajectories. If there is an additional constant of the motion, then the motion takes place on a surface. Under a perturbation only some of the irrational KAM surfaces are preserved, but these surfaces isolate regions in three-dimensional space. A system trajectory that starts in one region is prevented from crossing into another by the KAM surfaces preserved under the perturbation.

If we now consider a Hamiltonian system with three degrees of freedom, the "surfaces" on which the motion takes place are three-dimensional manifolds embedded in six-dimensional phase space. The KAM surfaces preserved under a perturbation do not necessarily isolate regions of three-dimensional configuration space from each other. Consequently, two arbitrarily selected regions in configuration space may be in a region that is confined between three-dimensional KAM surfaces. The additional dimensions allow widely separated points in configuration space to be connected by trajectories that go into the other dimensions. This interconnected region between preserved KAM surfaces is often referred to as the *Arnold web*.

Because these higher dimensional surfaces are hard to visualize an analogy can be helpful. Figure 8.9 shows how lines in two-dimensional space can isolate regions and trap trajectories within them. Yet in three-dimensional space these same lines do not isolate regions, and phase-space trajectories can wander throughout the region.

A way of displaying the Arnold web in a system with three degrees of freedom has been outlined by Martens et al. (1987). The basic idea is to display the diffusion through phase space of a perturbed Hamiltonian system by following the change with time in the action momenta associated with the unperturbed Hamiltonian H_0. If the perturbation were zero, then the values of these action momenta would remain constant, i.e., the system point would remain on an invariant torus with constant values for the action momenta. However, rather than actually following the action momenta, the frequencies

$\omega^i(\mathbf{J})$ are observed. This is equivalent as long as the relationship between the frequencies and the action momenta is invertible. One of the action momenta can be taken to be the Hamiltonian itself. Thus, in the system including the perturbation, the quantities to follow are two independent ratios of frequencies. Without the perturbation both of these ratios would remain constant. As Arnold diffusion takes place, the system point moves from one resonance zone to another. Consequently, a plot with the frequency ratios ω_1/ω_3 and ω_2/ω_3 along the x and y axes supplies a visualization of the Arnold web and is sometimes referred to as "tune space." For further detail and discussion we refer the reader to Martens et al. (1987).

It is not necessary, however, to go to a system with three degrees of freedom to have an Arnold web. This can already happen in $1\frac{1}{2}$ dimensions when the perturbation of the Hamiltonian system is resonant with the natural frequencies of the system (Chernikov et al. 1987). In this circumstance the regions of phase space where the KAM surfaces are preserved may be dramatically reduced. Color Plate IV shows a Poincaré section for a perturbed Hamiltonian system with an equation of motion

$$\ddot{x} + \omega_0^2 x = \epsilon \sin(kx - \Omega t). \tag{8.55}$$

This system serves as a model for particle motion in an external, uniform magnetic field in the presence of a plane wave that propagates perpendicular to the magnetic field. The resonance condition is established by choosing $\Omega = 5\omega_0$. Every color in Plate IV represents a different choice for the initial conditions. The possibility of diffusion throughout large regions of phase space by motion along the Arnold web is evident. The particle can gain or loose energy to the wave as it moves within the web.

8.7 Summary

We have seen in this chapter that it is only very special Hamiltonians that have a complete set of isolating integrals of the motion, so that the dynamical evolution takes place on invariant toroids called KAM surfaces. Perturbations to these Hamiltonians destroy some of these KAM surfaces and lead to regions of chaotic dynamics. Thus, the generic Hamiltonian system has regions in phase space (perhaps small) where the trajectories of the system are chaotic. These chaotic regions in phase space are a consequence of resonance overlap and their interconnection is referred to as the Arnold web.

Exercises

8.1 Compute S_1 of (8.43) if the perturbed Hamiltonian $H_1 = \frac{J^2}{6}\cos^4\theta$.

8.2 Explore numerically the area-preserving Hénon map of (7.44). Illustrate for a specific choice of the map parameter c the existence of good KAM surfaces and chaotic regions in the (x, y) plane.

8.3 Consider the differential equation

$$\ddot{x} + \omega_0^2 x = (K/k_0 T^2) \sin k_0 x \sum_{n=-\infty}^{\infty} \delta(t/T - n) \qquad (8.56)$$

for the behavior of a linear oscillator under the action of periodic kicks. Obtain a two-dimensional mapping by integrating this equation and explore the structure of the Arnold web as a function of the strength parameter K. For a report of results see Chernikov et al. (1987).

CHAPTER NINE

MEASURES OF CHAOS

*The order of nature, as perceived at first
glance, presents at every instant a chaos followed by another chaos. We must decompose
each chaos into single facts. We must learn
to see in the chaotic antecedent a multiple of
distinct antecedents, in the chaotic consequent
a multitude of distinct consequents.*
(J. S. Mill, Logic, bk III, Chap. VII)

In previous chapters we focused our attention on the chaotic dynamics of
nonlinear systems. We examined simple systems and used various methods
to characterize the chaos, including Poincaré sections, Lyapunov exponents,
and fractal dimension. In this chapter we direct our attention specifically to
the issue of measuring and characterizing chaos. We meet again some topics
introduced earlier, but consider them in greater depth and usually with an
eye toward experimental applications.

9.1 Power Spectrum Analysis

One frequently applied technique for understanding experimental data is
Fourier analysis. Fourier analysis is a method for "projecting out" information contained in a measured signal. Such analysis is useful for obtaining
among other things, the frequency components and the power distribution
as a function of frequency. In this section we give only a brief introduction
to typical results and standard methods of Fourier analysis. For more detail
we refer the reader to Bergé et al. (1984, Chapter 3) for a discussion in the
context of chaos. Many good books exist on signal analysis using Fourier
transforms and Fante (1988) is a recent exposition. For numerical techniques
and a concise discussion of background theory we recommend Press et al.
(1986). We also wish to point out that it is only because of the emergence
in the 1960s of the numerical technique known as the fast Fourier transform
that practical Fourier analysis is possible. For a serious implementation of
the ideas presented here and a discussion of the many subtleties associated
with the analysis of experimental data, we urge the reader to consult the
aforementioned (or equivalent) references.

Let $s(t)$ denote a signal that is a function of time, and let $s_\alpha, \alpha = 1, 2, \ldots, N$ denote a discrete set of measured values taken at the times $t_\alpha = \alpha \Delta t, \alpha = 1, 2, \ldots, N$. The quantity Δt represents the time interval between samples and is assumed to be uniform. The reciprocals of t_α represent frequencies and the *Nyquist critical frequency* $\nu_c = 1/2 \Delta t$ plays a pivotal role. If the frequency content of the signal $s(t)$ is limited to frequencies below ν_c, then a result known as the *sampling theorem* guarantees that the entire information content of $s(t)$ is captured by the finite set of sampled values $\{s_\alpha\}$. If the time interval between samples is too large, so that the effective bandwidth of the signal $s(t)$ does not lie below ν_c, then the reconstructed signal and the power spectrum are distorted by the components of $s(t)$ not represented by the s_α. This distortion is called *aliasing* and is recognized by the fact that the Fourier transform of $s(t)$, namely $\hat{s}(\nu)$, does not approach zero as $\nu \to \pm\nu_c$. There is little to be done if a given finite sample exhibits aliasing. But in advance one can filter the signal or take an adequately small Δt so that the sample rate is sufficiently high. A determination of the natural bandwidth of the signal makes possible such steps to ensure that aliasing is not a problem. Other forms of error encountered in finite samples are discussed in the references cited.

We now take the discrete Fourier transform of the finite data set $\{s_\alpha\}$. We write the elements of this set in the form

$$s_\alpha \equiv s(\alpha \, \Delta t) = \frac{1}{N} \sum_{\beta=1}^{N} \hat{s}_\beta e^{-i2\pi\beta(\alpha/N)}, \tag{9.1}$$

where the $\hat{s}_\beta$ is called the *discrete Fourier transform* of the N points $\{s_\alpha\}$. Using the fact that

$$\sum_{\alpha=1}^{N} [e^{-i2\pi\beta(\alpha/N)}][e^{-i2\pi\gamma(\alpha/N)}]^* = N\delta_{\beta\gamma}, \tag{9.2}$$

we find immediately that

$$\hat{s}_\alpha = \sum_{\beta=1}^{N} s_\beta e^{i2\pi\beta(\alpha/N)}. \tag{9.3}$$

Using (9.2) leads to Parseval's theorem for the discrete Fourier transform:

$$\sum_{\alpha=1}^{N} |s_\alpha|^2 = \frac{1}{N} \sum_{\alpha=1}^{N} |\hat{s}_\alpha|^2. \tag{9.4}$$

We now define the *autocorrelation function*:

$$C_\gamma = \sum_{\alpha=1}^{N} s_\alpha s_{\alpha+\gamma}. \tag{9.5}$$

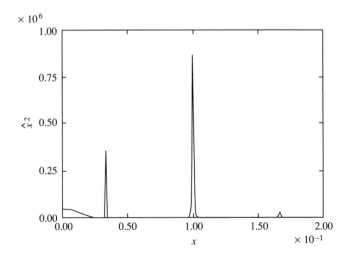

FIGURE 9.1   Power spectrum for a finite set of points along a trajectory for the forced, damped pendulum with parameters as in Fig. 6.11 where the attracting set is a period-3 limit cycle.

Note that in (9.5) the index $\alpha + \gamma$ can certainly exceed N. What is to be done with such indices? To this point we have ignored the symmetries in the discrete Fourier transformation. However, we remedy this situation now since it is important for the issue raised in (9.5) and also because the symmetries are useful for increasing the computational efficiency. From (9.1) and (9.3) we see immediately that

$$s_0 = s_N, \quad \hat{s}_0 = \hat{s}_N,$$
$$s_{\alpha-N} = s_{\alpha+N} = s_\alpha,$$
$$\hat{s}_{\alpha-N} = \hat{s}_{\alpha+N} = \hat{s}_\alpha. \tag{9.6}$$

We have tacitly assumed that the set of measured signal values $\{s_\alpha\}$ consists of real numbers. This implies for the the transform that

$$\hat{s}_\beta = \hat{s}^*_{N-\beta}. \tag{9.7}$$

Substituting from (9.1) into (9.5) and using the orthogonality relation (9.2), we find immediately that

$$C_\gamma = \frac{1}{N} \sum_{\beta=1}^{N} |\hat{s}_\beta|^2 e^{-i2\pi\beta(\gamma/N)}, \tag{9.8}$$

and the inverse of this relation

$$|\hat{s}_\beta|^2 = \sum_{\gamma=1}^{N} C_\gamma e^{i2\pi\gamma(\beta/N)}. \tag{9.9}$$

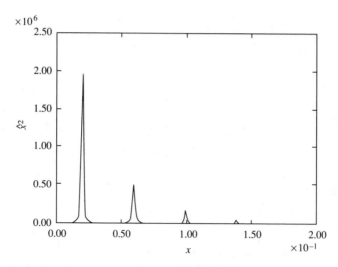

FIGURE 9.2   The same as Fig. 9.1 except for parameters as given in Fig. 6.12 yielding a period-5 limit cycle.

Relations (9.8) and (9.9) show that the autocorrelation function and the power in the βth frequency component form a discrete transform pair (Wiener–Khintchin theorem). The graph of $|\hat{s}_\beta|^2$ as a function of the frequency $\nu_\beta = \beta/(N\Delta t)$ is called the *power spectrum*.

As an example we apply these methods to the forced, damped pendulum described in Section 6.4. We sample the phase-space points at intervals of $\Delta t = T/10$, where T is the period of the forcing term in (6.28). We take

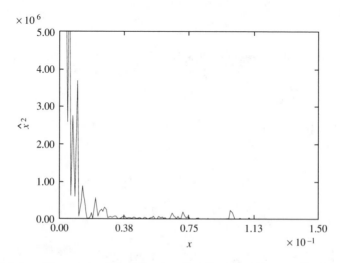

FIGURE 9.3   The same as Fig. 9.1 except that the parameters are for the completely chaotic state with Poincaré plot given in Plate III.

1024 successive values of the angular displacement $x(\Delta t), x(2\,\Delta t), \ldots$ as our finite data set $\{s_\alpha\}$. Using fast-Fourier transform numerical techniques, we find then the power spectrum corresponding to the given finite data set. Figures 9.1 to 9.3 represent the power spectrum for the three cases depicted in Figs. 6.10, 6.13, and 6.14, respectively. In all cases we see a peak at $\nu_0 \simeq 0.1$ corresponding to the fundamental drive frequency. For the 3-cycle there is a peak at $\nu_0/3$ and similarly at $\nu_0/5$ for the 5-cycle. For the case with chaos the power spectrum has typically noisy behavior at low frequencies. For a chaotic signal we expect all the power to be in the low frequencies. This is because aperiodic points in a finite data set appear as points with very long periods, comparable to the total sample time, and consequently, correspond to very low frequencies.

9.2 Lyapunov Characteristic Exponents Revisited

We have already introduced the concept of a Lyapunov characteristic exponent (LCE) in Section 2.2. However, in that section the discussion was restricted to one dimension. In this section we expand these results into more than one dimension and devote some attention to numerical considerations. We refer the interested reader to Lichtenberg and Lieberman (1983), Benettin and Galgani (1979), Benettin et al. (1976), and Shimada and Nageshima (1979) for additional reading. For mathematical questions such as existence proofs for limits, we refer the reader to Oseledec (1968).

Recall that for a dynamical system represented by a differential equation the trajectories are refered to as flows (see the introduction to Chapter 6). Let M denote the phase-space manifold of an arbitrary system and denote a flow in M as

$$\phi^t \colon M \to M. \tag{9.10}$$

The notation suggests that one takes an initial point $x_0 \in M$, and ϕ^t maps this initial point to $\phi^t(x_0) \equiv x(t)$. As t varies, one obtains the trajectory of the dynamical system passing through the initial point x_0. A flow is a one-parameter group of diffeomorphisms with the composition law

$$\phi^{t_2+t_1} = \phi^{t_2} \circ \phi^{t_1}. \tag{9.11}$$

We assume that there is some notion of distance defined on M (usually a Riemannian metric) and denote the distance between two points $x, y \in M$ as $d(x,y)$. The notion of distance is clearly important for an LCE since an LCE is designed to measure separation of neighboring trajectories under the flow. For a fixed time t we consider the ratio

$$d\big(\phi^t(x), \phi^t(y)\big)/d(x,y)$$

in the limit that $d(x,y) \to 0$. To quantify this limit we consider a regular curve $C(s)$ such that $C(0) = x$. See Fig. 9.4. Then the limit of the distance

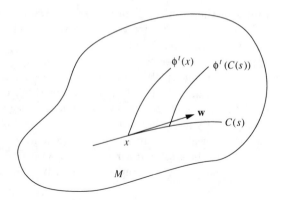

FIGURE 9.4   Neighboring trajectories in phase space.

$d(x, y)$ going to zero can be written in the form

$$\lim_{s \to 0} d\Big(\phi^t\big(C(s)\big), \phi^t(x)\Big) / d\big(C(s), x\big). \tag{9.12}$$

We could now take the logarithm of this expression, divide by t, and take the limit as $t \to \infty$ for the definition of LCE. However, this definition in terms of distances is less convenient than one using the vectors in the tangent space $T_x M$ at $x \in M$.

Let $\| \cdot \|$ denote the norm on the tangent spaces of M induced by the distance measure on M. Then $d\big(C(s), x\big) = s\|\mathbf{w}\| + o(s)$, where $\mathbf{w} = dC/ds(0) \in T_x M$ and $o(s)/s \to 0$ as $s \to 0$. Assuming t to be fixed, ϕ^t is a mapping (diffeomorphism) of M onto M. The linearization of this map at $x \in M$ is simply the familiar derivative map $D\phi_x^t$. The operator $D\phi_x^t$ linearly maps vectors in $T_x M$ to vectors in $T_{\phi^t(x)} M$. For completeness we give an explicit representation of this linear operator $D\phi_x^t$.

In terms of coordinates we write the arbitrary curve $C(s)$ as

$$C(s) = \big(x^1(s), x^2(s), \dots, x^n(s)\big), \tag{9.13}$$

where $C(0) = x = \big(x^1(0), x^2(0), \dots, x^n(0)\big)$. Refer to Fig. 9.4 for a sketch of the arbitrary curve $C(s)$ and the neighboring trajectories $\phi^t(x)$ and $\phi^t(C(s))$. At $x = C(0)$ the tangent vector to this curve is given by

$$\mathbf{w} = \frac{dC(s)}{ds}\bigg|_{s=0} \equiv \dot{C}(0). \tag{9.14}$$

Then the derivative map $D\phi_x^t$ on tangent vectors is defined by

$$D\phi_x^t[\mathbf{w}] = \frac{d}{ds}\Big(\phi^t\big(C(s)\big)\Big)\bigg|_{s=0}. \tag{9.15}$$

If we denote the image point in terms of coordinates

$$\phi^t(C(s)) = \left(\bar{x}^1(C(s)), \bar{x}^2(C(s)), \ldots, \bar{x}^n(C(s))\right),$$

then (9.15) gives an explicit representation of the linear map $D\phi_x^t$ in the form

$$(D\phi_x^t[\mathbf{w}])^j = \frac{\partial \bar{x}^j}{\partial x^i} \frac{dx^i}{ds}\Big|_{s=0} = \frac{\partial \bar{x}^j}{\partial x^i} w^i. \tag{9.16}$$

The composition rule (9.11) implies for $D\phi_x^t$ that

$$D\phi_x^{t_2+t_1}[\mathbf{w}] = D\phi_{\phi^{t_1}(x)}^{t_2} \circ D\phi_x^{t_1}[\mathbf{w}]. \tag{9.17}$$

Returning to (9.12) we can now write

$$\lim_{s \to 0} \frac{s\|D\phi_x^t[\mathbf{w}]\| + o(s)}{s\|\mathbf{w}\| + o(s)} = \frac{\|D\phi_x^t[\mathbf{w}]\|}{\|\mathbf{w}\|}. \tag{9.18}$$

Now define the Lyapunov characteristic exponent (LCE) of $\mathbf{w} \in T_x M$ to be

$$\lambda(x, \mathbf{w}) = \lim_{t \to \infty} \frac{1}{t} \ln\left\{\frac{\|D\phi_x^t[\mathbf{w}]\|}{\|\mathbf{w}\|}\right\}. \tag{9.19}$$

With the composition law (9.17) we see that (9.19) is consistent with the one-dimensional result (2.13). Since the limit is being taken as $t \to \infty$, the denominator in (9.19) can be discarded, and we may write

$$\lambda(x, \mathbf{w}) = \lim_{t \to \infty} \frac{1}{t} \ln\|D\phi_x^t[\mathbf{w}]\|. \tag{9.20}$$

For a map

$$x_{n+1} = f(x_n) \tag{9.21}$$

instead of a differential flow, the definition (9.20) becomes

$$\lambda(x, \mathbf{w}) = \lim_{n \to \infty} \frac{1}{n} \ln\|(Df(x))^n[\mathbf{w}]\|, \tag{9.22}$$

where

$$(Df(x))^n = Df(f^n(x)) \circ \cdots \circ Df(x)[\mathbf{w}]. \tag{9.23}$$

In order to solidify these ideas consider a simple example. Let the point x be on a periodic orbit, i.e., $\phi^\tau(x) = x$, where τ is the period. In this circumstance $D\phi_x^\tau$ is a linear operator on $T_x M$. Suppose that the dimension of M is n and that the eigenvalues of $D\phi_x^\tau$ satisfy $|\lambda_1| > |\lambda_2| > \cdots > |\lambda_n|$ with $e_1, e_2, \ldots, e_n$ denoting the corresponding eigenvectors. For any $j > 0$ and an i, $1 \le i \le n$ it follows that

$$D\phi_x^{j\tau}[e_i] = (\lambda_i)^j e_i. \tag{9.24}$$

By letting $t = j\tau$ we obtain from (9.20)

$$\lambda(x, e_i) = \lim_{j \to \infty} \frac{1}{j\tau} \ln \|(\lambda_i)^j e_i\| = \frac{1}{\tau} \ln |\lambda_i|. \tag{9.25}$$

For an arbitrary vector $\mathbf{w}$ that can be expressed as a linear combination of the eigenvectors, it is straightforward to show that (Exercise 9.2)

$$\lambda(x, \mathbf{w}) = \frac{1}{\tau} \ln |\lambda_1|. \tag{9.26}$$

In other words, for an arbitrary $\mathbf{w}$, the largest eigenvalue rapidly dominates.

Even in the general case when eigenvectors and eigenvalues are not available, the largest LCE rapidly dominates as $t \to \infty$. We turn our attention now to the numerical computation of the largest LCE.

In using (9.20) to compute a LCE it is necessary to know how the flow propagates an arbitrary vector $\mathbf{w}$ forward along a trajectory. To this end we make a slight change in notation for (9.16). We denote $D\phi_x^t[\mathbf{w}]$ as $\mathbf{w}(t)$, with $\mathbf{w}(0) = \mathbf{w}_0 \equiv \mathbf{w}$. Then (9.16) may be written in the form

$$w^j(t) = \frac{\partial \bar{x}^j(t)}{\partial x^i} w_0^i. \tag{9.27}$$

Differentiating (9.27) with respect to time we find

$$\frac{dw^j(t)}{dt} = \frac{\partial}{\partial x^i} \left(\frac{d\bar{x}^j}{dt} \right) w_0^j. \tag{9.28}$$

The flow is defined by a differential equation in the usual way

$$\dot{x}^j = f^j(x, t). \tag{9.29}$$

Substituting (9.29) into (9.28) we find

$$\frac{dw^j(t)}{dt} = w_0^i \frac{\partial f^j}{\partial x^i}(\bar{x}, t) = w_0^i \frac{\partial \bar{x}^k}{\partial x^i} \frac{\partial f^j}{\partial \bar{x}^k}(\bar{x}, t) = w^k \frac{\partial f^j}{\partial \bar{x}^k}(\bar{x}, t). \tag{9.30}$$

In vector notation (9.30) takes the compact form

$$\dot{\mathbf{w}}(t) = \mathbf{w} \cdot \nabla \mathbf{f}. \tag{9.31}$$

This is our new flow equation for determining the LCE. Equation (9.20) for the LCE in this notation takes the transparent form

$$\lambda(x, \mathbf{w}_0) = \lim_{t \to \infty} \frac{1}{t} \ln \|\mathbf{w}(t)\|. \tag{9.32}$$

Taking the limit $t \to \infty$ in (9.32) in naive fashion promptly leads to computer overflow. We can circumvent this problem as follows: Choose an initial $\mathbf{w}_0$ with unit norm, $\|\mathbf{w}_0\| = 1$. Evolve $\mathbf{w}$ to a point in time t_1 using (9.31) and then let

$$\mathbf{w}_1(t) \equiv \mathbf{w}(t)/\|\mathbf{w}(t_1)\| = \mathbf{w}(t)/\alpha_1. \tag{9.33}$$

Then $\mathbf{w}_1 = \mathbf{w}/\alpha_1$ and

$$\dot{\mathbf{w}}_1 = \frac{1}{\alpha_1}\dot{\mathbf{w}} = \frac{1}{\alpha_1}\mathbf{w} \cdot \nabla \mathbf{f} = \mathbf{w}_1 \cdot \nabla f. \tag{9.34}$$

We repeat this process and evolve $\mathbf{w}_1$ to a time t_2. Define $\mathbf{w}_2(t) = \mathbf{w}_1(t)/\alpha_2$, where $\alpha_2 \equiv \|\mathbf{w}_1(t_2)\|$ and then $\|\mathbf{w}(t_2)\| = \alpha_1\alpha_2$. Continuing this process for n steps yields

$$\|\mathbf{w}(t_n)\| = \prod_{i=1}^{n} \alpha_i. \tag{9.35}$$

Then

$$\lambda(x, \mathbf{w}_0) = \lim_{n \to \infty} \frac{1}{t_n} \ln \left(\prod_{i=1}^{n} \alpha_n \right) = \lim_{n \to \infty} \frac{1}{t_n} \sum_{i=1}^{n} (\ln \alpha_i). \tag{9.36}$$

If $t_n = n\Delta t$, i.e., if we use a uniform step size for advancing the vector $\mathbf{w}(t)$, then (9.36) becomes

$$\lambda(x, \mathbf{w}_0) = \lim_{n \to \infty} \frac{1}{n} \sum_{i=1}^{n} (\ln \alpha_i)/\Delta t. \tag{9.37}$$

It is clear that the LCE depends on the vector $\mathbf{w}_0$. If for example $\mathbf{w}_0$ were chosen at x to be the tangent vector to the trajectory, then there would be no divergence and the LCE would be zero. Similarly, in the example considered earlier for a periodic orbit, if $\mathbf{w}_0$ were chosen exactly equal to an eigenvector then the LCE is just the logarithm of the corresponding eigenvalue, divided by the period. In the general case, however, where the initial vector $\mathbf{w}_0$ is chosen at random, $\mathbf{w}_0$ will have a component in every subspace. Thus we expect that for an arbitrarily chosen vector $\mathbf{w}_0$ in the tangent space $T_x M$, (9.37) leads to the largest LCE. If this LCE is positive, then we have a strong indicator of chaotic dynamics.

In Fig. 9.5 the LCEs for the driven, damped pendulum are plotted with parameter values corresponding to those for Fig. 6.14 and Fig. 9.3. We see in this figure that the largest LCE is positive, indicating chaos. This is in agreement with the Poincaré section and the power spectrum for this system.

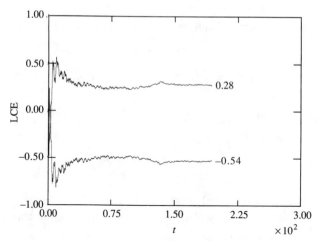

FIGURE 9.5   The LCEs for a driven, damped pendulum with parameters that correspond to chaotic motion as in Figures 6.14 and 9.3.

Let us now consider the other LCEs. It follows immediately from the triangle inequality that

$$\lambda(x, \mathbf{w} + \mathbf{w}') \leq \max[\lambda(x, \mathbf{w}), \lambda(x, \mathbf{w}')]. \tag{9.38}$$

It is also straightforward to show that

$$\lambda(x, c\mathbf{w}) = \lambda(x, \mathbf{w}). \tag{9.39}$$

With the results (9.38), (9.39), and the definition that $\lambda(x, 0) = -\infty$, it is a simple exercise to show that

$$L_a = \{\mathbf{w} \in T_x M | \lambda(x, \mathbf{w}) \leq a\} \tag{9.40}$$

is a linear subspace of $T_x M$. If the dimension of M is n, then it is clear that there can be at most n distinct values $\lambda_1, \ldots, \lambda_s$, $1 \leq s \leq n$ such that $\lambda(x, \mathbf{w}) = \lambda_i$. In general both the number s and the numbers $\lambda_1, \ldots, \lambda_s$ depend on the point $x \in M$. We now define a set of linear subspaces of $T_x M$.

$$L_i(x) = \{\mathbf{w} \in T_x M | \lambda(x, \mathbf{w}) \leq \lambda_i\} \quad \text{and} \quad L_{s+1}(x) = \{0\}. \tag{9.41}$$

We then have what is called a *filtration* of $T_x M$.

$$T_x M = L_1(x) \supset L_2(x) \supset \cdots \supset L_{s+1} = \{0\}$$

with the basic property that $\lambda(x, \mathbf{w}) = \lambda_i$ if and only if $\mathbf{w} \in L_i(x)$ but $\mathbf{w} \notin L_{i+1}(x)$. We assume that the values λ_i have been ordered such that

$\lambda_1 > \lambda_2 > \cdots > \lambda_s$. The number $k_i = \dim L_i(x) - \dim L_{i+1}(x)$ is called the multiplicity of λ_i, and the λ_i are referred to as the spectrum of LCEs at $x \in M$.

Since the lower dimensional subspaces L_i have zero measure in $T_x M$, an arbitrarily chosen vector $\mathbf{w}_0$ will always have a component lying outside of $L_2, L_3, \ldots, L_{s+1}$, even if that component is generated by numerical noise. Consequently, (9.37) always leads to the largest LCE λ_1. So how can we compute the other LCEs?

To this end we generalize the definition of the LCE for a vector at a point, which is really a measure of length, to one for a volume. Let $\mathbf{e}_1, \ldots, \mathbf{e}_n \in T_x M$ be a set of n-linearly independent vectors in $T_x M$, which we may assume to be a basis. Then it is convenient to use the exterior product of vectors to define volumes. As a reference to the exterior product for vectors we refer the reader to Schutz (1980, Section 4.9). The exterior product of p basis vectors is denoted by

$$\mathbf{e}_{i_1} \wedge \cdots \wedge \mathbf{e}_{i_p}. \tag{9.42}$$

Such vectors are a generalization of the familiar cross product of vectors in three dimensions. They are referred to as p-vectors and form a vector space of dimension $(n!/p!(n-p)!)$. Geometrically, one can think of a p-vector as representing a p-dimensional parallelopiped. The representation of a p-dimensional volume in terms of p-vectors explains our use of p-vectors in the context of a generalized LCE, to be defined shortly.

As a specific example let $n = 4$ and $p = 3$, then the space of 3-vectors has a basis

$$\{\mathbf{e}_1 \wedge \mathbf{e}_2 \wedge \mathbf{e}_3, \; \mathbf{e}_1 \wedge \mathbf{e}_2 \wedge \mathbf{e}_4, \; \mathbf{e}_1 \wedge \mathbf{e}_3 \wedge \mathbf{e}_4, \; \mathbf{e}_2 \wedge \mathbf{e}_3 \wedge \mathbf{e}_4\}. \tag{9.43}$$

If $\mathbf{u}, \mathbf{v}, \mathbf{w}$ are three linearly independent vectors in $T_x M$, then $\mathbf{u} \wedge \mathbf{v} \wedge \mathbf{w}$ is a nonzero 3-vector. In the example with $n{=}4$ this 3-vector has components in the basis of (9.43)

$$\left\{ \sum_\pi \delta_\pi u^1 v^2 w^3, \; \sum_\pi \delta_\pi u^1 v^2 w^4, \; \sum_\pi \delta_\pi u^1 v^3 w^4, \; \sum_\pi \delta_\pi u^2 v^3 w^4 \right\}, \tag{9.44}$$

where the sum is over the permutations of the three indices and $\delta_\pi = +1 \, (-1)$ for an even (odd) permutation A norm on such p-vectors is a measure of the "volume" spanned by the p-dimensional parallelopiped defined by the p-vectors.

$$V_p = \|\mathbf{w}_{i_1} \wedge \cdots \wedge \mathbf{w}_{i_p}\|. \tag{9.45}$$

We know that the flow ϕ^t will distort such volumes. The expansion (contraction) for the different directions in general varies, which is a reflection of the LCE spectrum of the dynamical system. Consequently, we define a generalized LCE:

$$\lambda(x, E^p) = \lim_{t \to \infty} \frac{1}{t} \ln \|D\phi_x^t(E^p)\|, \qquad (9.46)$$

where E^p is spanned by $\mathbf{w}_{i_1}, \ldots, \mathbf{w}_{i_p}$ and

$$\|D\phi_x^t(E^p)\| \equiv \|D\phi_x^t(\mathbf{w}_{i_1}) \wedge \cdots \wedge D\phi_x^t(\mathbf{w}_{i_p})\|. \qquad (9.47)$$

The quantity $\|D\phi_x^t(E^p)\|$ defined in (9.47) denotes the propagated volume. Benettin et al. (1980) show that one can find p linearly independent vectors $\mathbf{w}_1, \ldots, \mathbf{w}_p \in E^p$ such that

$$\lambda(x, E^p) = \lambda(x, \mathbf{w}_1) + \cdots + \lambda(x, \mathbf{w}_p). \qquad (9.48)$$

Thus we see that each LCE of order p is the sum of p LCEs of order 1. And just as for the LCEs of order 1, for an E^p chosen at random, one obtains the sum of the p largest LCEs.

So now the method for computing all LCEs becomes clear. We compute generalized LCEs of order 1 through n. Then peel off one by one the individual LCEs using (9.48). There is, however, one additional difficulty in the computation besides the necessity to repeatedly renormalize. Because of the differential stretching in the various directions, all vectors tend to become aligned along the direction of maximum stretching, the one with the largest LCE. Thus one loses the linear independence of the n vectors after a short time. Hence, at each time step we must not only renormalize, but we must reorthogonalize the vectors being propagated. For this we use the familiar Gram–Schmidt procedure.

Let $\{\mathbf{w}_1(0), \ldots, \mathbf{w}_n(0)\}$ denote a set of orthonormal basis vectors for the tangent space at the point $x(0)$. We consider a propagation of these vectors along the trajectory $x(t)$ generated by the flow ϕ^t given by the differential equation (9.29). We integrate this flow in time, stopping at the times $0 < \Delta t < 2\Delta t < \cdots < m\Delta t$ to renormalize and reorthogonalize. In the time interval $j\Delta t < t < k\Delta t$

$$\mathbf{w}_1^{(j)}(t) = \mathbf{w}_1^{(j-1)}(t)/\alpha_1^{(j)}, \qquad (9.49)$$

where

$$\alpha_1^{(j)} = \|\mathbf{w}_1^{(j-1)}(j\Delta t)\|. \qquad (9.50)$$

Similarly the volumes are given by

$$V_1^{(1)} = \alpha_1^{(1)}, \; V_1^{(2)} = \alpha_1^{(1)}\alpha_1^{(2)}, \dots, V_1^{(m)} = \prod_{i=1}^{m} \alpha_1^{(i)}, \qquad (9.51)$$

with

$$\lambda(x, E^1) = \lim_{n\to\infty} \frac{1}{n} \sum_{i=1}^{n} \ln \alpha_1^{(i)}/\Delta t. \qquad (9.52)$$

This result is, of course, nothing more than (9.37).

For the higher dimensional volumes we need the other vectors as well. In the time interval $j\Delta t < t < k\Delta t$ and for $2 \le l \le n$

$$\mathbf{w}_l^{(j)} = \left[\mathbf{w}_l^{(j-1)} - \sum_{i=1}^{l-1} \left(\mathbf{w}_l^{(j-1)}(j\Delta t) \cdot \mathbf{w}_i^{(j)}(j\Delta t) \right) \mathbf{w}_i^{(j)}(t) \right] \Big/ \alpha_l^{(j)}, \qquad (9.53)$$

where

$$[\alpha_l^{(j)}]^2 = \|\mathbf{w}_l^{(j-1)}(j\Delta t)\|^2 - \sum_{i=1}^{l-1} \left(\mathbf{w}_l^{(j-1)}(j\Delta t) \cdot \mathbf{w}_i^{(j)}(j\Delta t) \right)^2. \qquad (9.54)$$

From (9.53) and (9.54) follows

$$V_l^{(m)} = \prod_{i=1}^{m} \prod_{j=1}^{l} \alpha_j^{(i)}, \qquad (9.55)$$

and then

$$\lambda(x, E^l) = \lim_{n\to\infty} \frac{1}{n} \sum_{i=1}^{n} \sum_{j=1}^{l} \ln \alpha_j^{(i)}/\Delta t$$

$$= \sum_{j=1}^{l} \left[\lim_{n\to\infty} \frac{1}{n} \sum_{i=1}^{n} \ln \alpha_j^{(i)}/\Delta t \right]. \qquad (9.56)$$

By successively peeling off the LCEs, we obtain from (9.56)

$$\lambda(x, \mathbf{w}_l) = \lim_{n\to\infty} \frac{1}{n} \sum_{i=1}^{n} \ln \alpha_l^{(i)}/\Delta t. \qquad (9.57)$$

In this way we can compute all LCEs for a dynamical system.

One important reason for obtaining all the LCEs is to compute the Lyapunov fractal dimension. In Section 3.3 we discussed the Lyapunov dimension d_L and now that a method is available for computing all LCEs, it is a simple matter to obtain the Lyapunov fractal dimension. Figure 9.5 shows the result of the computation of the LCEs for a driven, damped pendulum with parameters corresponding to chaotic motion. From (3.77) we obtain a fractal dimension for this chaotic attractor $d_L \simeq 1.52$. Recall that the negative LCEs contract volumes leading in this case to a fractal dimension < 2.

9.3 Information and K-Entropy

As we have seen, LCEs provide an important measure of the chaotic nature of dynamical systems. This measure can be further quantified by considering the relationship to information content. Following a common introduction to information measure, we consider the number of possible states of a system represented by squares in a rectangular grid.

Table 9.1: Information		
	Number of states=Ω	Number of questions $= I$
	2	1
	4	2
	8	3
	16	4

By refering to Table 9.1 we see that the amount (bits) of information obtained by ascertaining which of the possible states of a system is realized is given by the relation

$$I = \lg \Omega, \tag{9.58}$$

where $\lg$ denotes the logarithm to the base 2 and Ω measures the number of possible states. If $\Omega_i = 2^{n-1}$ and $\Omega_f = 2^n$, then $\Delta I = \lg \Omega_f - \lg \Omega_i = 1$. In other words the information needed to specify the state of the system has increased by one bit.

The general definition of information is given in terms of the probabilities of the various states of the system (Shannon and Weaver, 1949). Assuming that p_i denotes the probability that the system is in the ith state, then

$$I = -\sum_i p_i \lg p_i. \tag{9.59}$$

Suppose the system has Ω states and each is equally probable, then $p_i = 1/\Omega$, independent of i, and we see that (9.59) reduces to (9.58).

We can now relate these concepts to LCEs by noting for a classical dynamical system that the number of states is essentially proportional to the volume in phase space. Suppose that the minimum uncertainty in the measurement of a phase-space variable is denoted as ϵ. Then the number of initial states Ω_0 satisfies

$$\Omega_0 \propto V_p(0)/\epsilon^n = \left(\frac{l_1}{\epsilon}\right) \times \cdots \times \left(\frac{l_n}{\epsilon}\right), \tag{9.60}$$

where l_i denotes the length of the side of the volume $V_p(0)$ in the ith direction. As the system evolves (9.46) and (9.48) imply that

$$V_p(t) = V_p(0)e^{t\lambda(x,E^p)} = V_p(0)e^t\sum \lambda_i. \tag{9.61}$$

For a conservative system we know that phase-space volume is preserved under the flow (Liouville's theorem) and thus

$$\sum_{i=1}^n \lambda_i = 0 \qquad \text{(conservative system)}. \tag{9.62}$$

Referring to (6.26) we see that the sum of the LCEs is equal to the divergence of the vector field determining the flow. For dissipative systems

$$\sum_{i=1}^n \lambda_i < 0 \qquad \text{(dissipative system)}. \tag{9.63}$$

As the system evolves, the number of states also evolves. However, due to the finite uncertainty no factor in (9.60) can become less than 1. Consequently,

$$\Omega(t) \propto \exp\left\{ t \sum{}' \lambda_i \right\}, \tag{9.64}$$

and

$$\frac{dI}{dt} = \sum{}' \lambda_i, \tag{9.65}$$

where the prime on the sums denotes that the sum is only over the positive LCEs. We have further rescaled the LCEs so that they are measured in units of bits per unit time.

One thing we have ignored is the possible dependence of the generalized LCE occurring in (9.61) on the initial point. This dependence of the LCEs on the position is included in the usual definition of the Kolmogorov-entropy, or K-entropy for brevity.

$$K = \int_M \rho(x) \sum{}' \lambda_i(x)\, dx, \tag{9.66}$$

where $\rho(x)$ represents an invariant distribution function, normalized to unity, on the attractor M (Benettin et al., 1976). Most frequently systems are considered that have LCEs independent of the point on the attractor and thus because the distribution function is normalized, the K-entropy is given by

$$K = \sum{}' \lambda_i. \tag{9.67}$$

The K-entropy represents the rate of information production and the growth of uncertainty. In a chaotic system with $K > 0$ the number of possible states of the system that evolve from some initial distribution increases with time. Hence information is increasing. Another way of viewing this information growth focuses on making a measurement of the system point in phase space. In a system with $K > 0$, a determination of the system point to within a minimum uncertainty at time $t > 0$ determines the initial state of the system to within an uncertainty at $t = 0$ much less than the minimum. Consequently, there has been a growth in information. As a chaotic system evolves the subsequent knowledge about the state of the system, as determined by the initial state, deteriorates. The system is generating information and unless a subsequent measurement is performed, one knows less and less about the system.

The K-entropy is the rate of information production and hence provides a natural measure of the time scale over which knowledge about the state of a system decays. For the driven, damped pendulum with parameters corresponding to Fig. 9.5, the time scale for information decay $\tau \sim 1/\lambda_1 \sim 4$ sec.

These relationships between K-entropy, information, and LCEs also suggest a heuristic derivation of the Lyapunov fractal dimension (4.13). As before, let k be the index denoting the last positive LCE and let n denote the phase-space dimension of the dynamical system under study. Certainly k of the n dimensions do not collapse under the dynamical flow because of the positive LCEs. To this remaining integer dimension k, we add a fractional piece that is computed by (4.4).

$$d_L = k + \lim_{\epsilon \to 0} \frac{K \Delta t}{\lg (1/\epsilon)}. \tag{9.68}$$

The information I is approximated as $K \Delta t$, where given ϵ, Δt is the smallest meaningful time increment, i.e., $\epsilon = 2^{\lambda_{k+1} \Delta t}$. For any interval longer than this, the reduction in phase volume is not measurable. Substituting this ϵ in (9.68) gives

$$d_L = k - \frac{\sum_{i=1}^{k} \lambda_i}{\lambda_{k+1}},$$

which is just (4.13).

9.4 Generalized Information, K-Entropy, and Fractal Dimension

The Shannon information used in the previous section is familiar, intuitive, and its rate of change is related to the K-entropy and LCEs. This notion of information can be generalized, however, in a way that unifies many of the measures of chaos that have been previously introduced. For further discussion of generalized information we refer the interested reader to Atmanspacher and Morfill (1986), Grassberger (1985), and Hentschel and Procaccia (1983).

For the real number $\alpha > 0$, $\alpha \neq 1$ we define the generalized information to be

$$I_\epsilon^{(\alpha)} = \frac{1}{1-\alpha} \lg \sum_{i=1}^{n} p_i^\alpha, \qquad (9.69)$$

where as before ϵ denotes the uncertainty in a phase-space measurement or the length of a cell side for a state-space partition. It requires only a careful application of L'Hospital's rule to show that the Shannon information (9.59) is given by $\lim_{\alpha \to 1} I_\epsilon^{(\alpha)} \equiv I_\epsilon^{(1)}$ (Exercise 9.6).

Using the generalized information of (9.69), we now define a generalized fractal dimension. We call this the generalized information dimension $d_I^{(\alpha)}$. Consistent with (3.72) we define

$$d_I^{(\alpha)} = \lim_{\epsilon \to 0} \frac{I_\epsilon^{(\alpha)}}{\lg (1/\epsilon)} = \lim_{\epsilon \to 0} \frac{1}{1-\alpha} \frac{\lg \left(\sum_i p_i^\alpha \right)}{\lg (1/\epsilon)}. \qquad (9.70)$$

From the fact that $\lim_{\alpha \to 1} I_\epsilon^{(\alpha)} = I_\epsilon^{(1)}$, it is clear that $d_I^{(1)}$ is just the information dimension of (4.4). Furthermore for $\alpha = 0$ we see that

$$d_I^{(0)} = \lim_{\epsilon \to 0} \frac{\lg (N)}{\lg (1/\epsilon)}, \qquad (9.71)$$

which is just the capacity dimension of (4.3). A somewhat less obvious, but equally satisfying result is that

$$d_I^{(2)} = \lim_{\epsilon \to 0} \frac{\lg \left(\sum_i p_i^2 \right)}{\lg \epsilon} = d_{\text{corr}}. \qquad (9.72)$$

This result follows by noting that, to within a normalization constant, the two-point correlation function $C(\epsilon)$ of (4.6) satisfies

$$C(\epsilon) = \sum_{i=1}^{n} p_i^2. \qquad (9.73)$$

In a similar way $d_I^{(3)}, \ldots$ are related to three-point correlation function, etc. Thus we see that all measures introduced in Chapter 4 for fractal dimension are included in the definition of generalized fractal dimension $d_I^{(\alpha)}$, except for the Lyapunov dimension d_L. Even d_L seems related to information as discussed in the previous section.

In a similar way generalized information can be used to obtain generalized measures of K-entropy.

$$K^{(\alpha)} \Delta t = \Delta I_\epsilon^{(\alpha)}, \tag{9.74}$$

where $\Delta I_\epsilon^{(\alpha)}$ represents the change in information of order α in the time interval Δt.

The relative sizes of the fractal dimensions $d_I^{(\alpha)}$ as α varies can also be determined. Straightforward differentiation gives

$$\frac{dI_\epsilon^{(\alpha)}}{d\alpha} = -\frac{1}{(1-\alpha)^2} \sum_{i=1}^{n} z_i \lg (z_i/p_i), \tag{9.75}$$

where $z_i = p_i^\alpha / \sum_i p_i^\alpha$. The sum occurring in (9.75) is called the Kullback information and is shown in Schlögl (1980) to be nonnegative. Consequently, we see that $dI_\epsilon^{(\alpha)}/d\alpha < 0$.

With this information that $I_\epsilon^{(\alpha)}$ is a monotonically decreasing function of α, we conclude

$$d_I^{(\alpha)} < d_I^{(\alpha')}, \qquad \alpha > \alpha'. \tag{9.76}$$

In particular we have

$$d_I^{(2)} = d_{\text{corr}} \leq d_I^{(1)} = d_I \leq d_I^{(0)} = d_{\text{cap}}, \tag{9.77}$$

consistent with the relation among these dimensions in (4.9). Equality in (9.77) is again only attained when the probability for each cell is the same and just given by $1/N$. A similar inequality holds for the generalized measures of information flow $K^{(\alpha)}$.

9.5 Chaos Measures from a Time Series

The problem of determining from experimental measurements quantities such as LCEs or fractal dimensions is entirely different from determining these quantities in a mathematical (numerical) investigation. Seldom is it the case that we even know what the correct phase-space variables are, let alone measure them all. Usually it is not even known how many variables are needed to fully describe the dynamics of the system. There is fortunately a partial answer to this problem that has been applied successfully in a large number of experimental investigations. The standard references on this topic are Packard et al. (1980) and Takens (1981).

The fundamental idea is easy to illustrate in the simple case of a nonlinear oscillator. In a system like the driven, damped pendulum discussed in Section 5.3, the fundamental phase-space variables are x and $\dot{x}$. To follow an evolution numerically, we follow x and $\dot{x}$ as functions of t. But since $\dot{x} = [x(t + \Delta t) - x(t)]/\Delta t$ in the limit as $\Delta t \to 0$, a knowledge of $x(t + \Delta t)$ is equivalent to a knowledge of $\dot{x}$. In other words a knowledge of a trajectory of points $[x(t), x(t + \Delta t)]$ is equivalent to a knowledge of the trajectory of points $[x(t), \dot{x}(t)]$.

This is the key idea: A phase-space trajectory:

$$\mathbf{x}(t) = [x_1(t), x_2(t), \ldots, x_n(t)]$$

is replaced by a trajectory in an artificial phase space with points given by $\mathbf{y}(t) = [y(t), y(t+\Delta t), \ldots, y(t+m\Delta t)]$, where $y(t)$ is any one of the phase-space variables $x_i(t)$ or perhaps even a functional combination of these variables. Thus from a set of measurements of a *single* quantity $y(t)$ we can construct a sequence of points in an artificial phase space:

$$\mathbf{x}(t) = [y(t), y(t + \Delta t), \ldots, y(t + m\Delta t)]$$
$$\mathbf{x}(t + \Delta t) = [y(t + \Delta t), \ldots, y(t + (m + 1)\Delta t)] \tag{9.78}$$
$$\vdots$$

Then from this series of points in the artificial phase space a correlation dimension can be calculated using (4.6), (9.72), and (9.73).

The choice of Δt is usually made after some experimentation. If Δt is too small then the $y(t), y(t + \Delta t), \ldots$ are not linearly independent. These data would also be linearly dependent if the system were periodic. If Δt is taken too large, i.e., much larger than the information decay time, then there is no dynamical correlation between the points. Often some trial and error is needed to get an appropriate Δt.

The second issue that must be faced in constructing the sequence of points (9.78) is how the number m is to be chosen. Usually the correlation dimension d_{corr} is computed for a series of values $m = 1, 2, \ldots$. The correlation dimension so computed increases with increasing m. However, once the dimension of the artificial phase space is large enough, the correlation dimension saturates and becomes constant. If m_0 is the minimum value of m for which d_{corr} approximately equals the saturated value, then $d = m_0 + 1$ is referred to as the *embedding dimension*. The quantity d represents the minimum dimensionality of the artificial phase space necessary to include the attractor. Subtleties and refinements in computing d_{corr} are discussed in Albano et al. (1988) and should be considered in research applications.

Figure 9.6 shows a plot of the numerator in (9.72) against the denominator for increasing values of m using data from a multimode laser, as reported by Atmanspacher and Scheingraber (1986). Figure 9.7 shows that the slope of these curves saturates at a value $d_{\text{corr}} \simeq 2.66$.

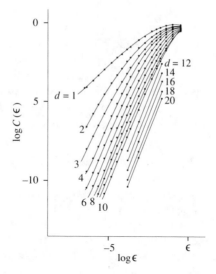

FIGURE 9.6   Log-log plot of the correlation integral $C(\epsilon)$ vs ϵ for increasing values of the artifical phase-space dimension. From Atmanspacher and Scheingraber (1986).

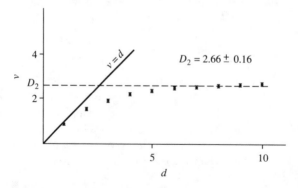

FIGURE 9.7   Change in the slope of the curves in Fig. 9.6 as a function of the embedding dimension m showing the convergence of $D_2 = d_I^{(2)} = d_{\text{corr}}$. From Atmanspacher and Scheingraber (1986).

9.6 Summary

In this chapter we have brought together some of the most important measures of chaos and indicated in some detail methods for their computation. The only tools not mentioned here were trajectory plots and Poincaré sections, however; these were used extensively in previous chapters. For any given system it is important to apply as many of these tools as possible in order that a convincing description of the chaotic behavior can be given. Chaos is sufficiently complex that no single tool is adequate for opening this complexity to our understanding.

Exercises

9.1 Verify Equation (9.7).

9.2 Show that (9.26) holds when $\mathbf{w} = c^i \mathbf{e}_i$.

9.3 Show for a Hamiltonian system that the LCEs come in pairs and satisfy the relations:

$$\sum_{i=1}^{2n} \lambda_i(x) = 0$$

$$\lambda_i(x) = -\lambda_{2n-i+1}(x).$$

9.4 Show that $\lambda_i(x)$ is a constant of the motion.

9.5 Supply the details for obtaining (9.55).

9.6 Show that $\lim_{\alpha \to 1} I_\epsilon^{(\alpha)} = -\sum_{i=1}^{n} p_i \lg p_i$.

CHAPTER TEN

COMPLEXITY AND CHAOS

Chaos *umpire sits, and by decision more im-*
broils the fray by which he reigns: next him
high arbiter Chance *governs all.*
(J. Milton, Paradise Lost, Bk II, 907-910)

One message made repeatedly clear by the chaotic, nonlinear systems studied in previous chapters is that deterministic chaos is not the same as randomness. Yet when presented with the binary string

$$(0110010100111101 0010110010) \tag{10.1}$$

one might well ask whether or not the string represents a random sequence of 0's and 1's. If one knew the source of the string, an answer could be given. Suppose the sequence (10.1) were generated by tosses of a coin with 1 denoting heads and 0 denoting tails. Then we would necessarily conclude that (10.1) is a random string since it is generated by a known random process. Suppose on the other hand that the binary string (10.1) were generated by the logistic map with $\mu < 4$ in the following way: A 0 is entered into the string if the image under the map falls to the left of the maximum, and the digit 1 is entered if the image falls to the right. This process we know to be deterministic, not random, and therefore we would conclude that the binary string (10.1) is not random.

The challenge comes when presented with a digit string such as (10.1) without any knowledge whatsoever as to the manner by which the string is generated. The digit string

$$(0101010101010101 01010101) \tag{10.2}$$

does not look as random as (10.1), but it is possible (although not probable) that this string was also generated by a random process such as coin tossing. However, the repeated pattern in (10.2) suggests that this binary string is not random.

Presented with a raw digit string such as (10.1) or (10.2), without any knowledge as to the process that generated the string, is it possible to decide whether the string is random or not? The somewhat surprising answer to this question is in general *no*. Furthermore, the undecidability of this question is not because of ineptitude but reflects a fundamental limitation rooted in the foundations of mathematics.

In the following section a definition of randomness is given that does not depend on any knowledge of the source of the string but depends only on the characteristics of the digit sequence; it is called *algorithmic complexity*. The definition is based on a concept of information similar to that introduced in the previous chapter. Our references on algorithmic complexity are Chaitin (1974; 1975). The measure of complexity used in subsequent sections and its connection to chaos are discussed by Kaspar and Schuster (1987), based on work of Lempel and Ziv (1976). Ford (1986; 1988) has discussed a role for algorithmic complexity in deciding fundamental issues in the foundations of physics.

10.1 Algorithmic Complexity

The algorithimic complexity of a string is defined to be the length in bits of the shortest algorithm required for a computer to produce the given string. For our purposes it is never necessary to assign an absolute value for the complexity of a string. Relative values are always sufficient. Consequently, the computer used in programming is not an issue since a program of finite complexity can always be constructed to relate one computer to another, or even one language on a computer to another.

The shortest algorithms are referred to as *minimal programs*. The complexity of a string is thus the length in bits of the minimal program necessary to produce the given string. The definition of a random number can now be given as any binary string whose algorithmic complexity is judged to be essentially equal to the length of the string. Qualitatively, the information embodied in a random number cannot be reduced or compressed to a more compact form.

Perhaps an example will help solidify these ideas. Consider the binary string (10.2). A simple computer program to produce the given string might be

$$\text{Print the string ``01'' 12 times.} \qquad (10.3)$$

As the number 12 is replaced with the number n, where n is large, the size of this program grows like $\lg n$.

On the other hand for the digit string (10.1) almost the best one can do is

$$\text{Print the string } S. \qquad (10.4)$$

where S denotes the string (10.1). As the string S grows in length, the length of program (10.4) grows like n, and the length of the computer program in bits is essentially the same as the length of S. Such a string satisfies the definition of a random number since the algorithmic complexity of the string is essentially the same as the length in bits of the string. Since the length of the string produced by program (10.3) is much longer in bits, $2n$ as compared to $\lg n$, the produced string is not random.

It can be proved that in a long, random digit string the relative frequency of appearance of 0 and 1 must be approximately equal to $\frac{1}{2}$. The strings like (10.2) demonstrate that equal relative frequency is not sufficient for a string to be random. However, within the set of all strings, those like (10.2) must be exceptional.

To study the occurrence of exceptional strings, let us consider binary strings of length n, where n is large, and count how many have a complexity less than $n - 10$. To measure the complexity we use a simple print program that is certainly not minimal. The program

Print the string "1" n times.

producing the string $(111\ldots1)$ has a complexity $c + 1$, where c is a constant coming from the programing statements and 1 comes from the simple string being repeated in this case. There are 2^1 strings of this type. There are 2^2 strings of the type (10.2) with the repeated string two digits long. Similarly there are 2^3 with three-digit repeated strings, etc. Adding all these with complexity less than $n - 10$, we have

$$2^1 + 2^2 + 2^3 + \cdots + 2^{n-11} = 2^{n-10} - 2.$$

Hence there are fewer than 2^{n-10} programs of size less than $n - 10$. Fewer than 2^{n-10} of the 2^n numbers have a complexity less than $n - 10$; $2^{n-10}/2^n = 1/1024$ and thus only about one number in 1000 has a complexity less than $n - 10$. In other words about one string in 1000 can be compressed into a program ten digits shorter than itself. By this complexity measure most n-digit strings are random.

We also remark briefly that this measure of complexity and definition of randomness has implications related to Gödel's incompleteness theorem. Since any formal system can in principle be encoded as a set of algorithms, the formal system has a finite length in bits. That is to say, any formal system has a finite information content. Thus the system is unable to decide on the truth or falsehood (randomness) of any statment that has as much information content (complexity) as the entire formal system. For further discussion of these very interesting ideas the reader is referred to the writings of Chaitin cited at the beginning of this chapter.

10.2 The LZ Complexity Measure

One of the major challenges of any researcher in chaotic dynamics is to extract a meaningful signal from data that have every appearance of being random. Clearly algorithmic complexity is a concept aimed specifically at the problem of distinguishing between the random and the nonrandom. The problem lies in determining a computable measure of complexity. No absolute measure is possible because minimal programs by definition correspond to random numbers, and it is not possible to determine a truly random number in any formal system. Nevertheless, it is possible to define a measure of complexity, and as we have seen earlier a relative measure is sufficient for many purposes. Our main references are Kasper and Schuster (1987), where these ideas are applied to dynamic systems exhibiting chaos, and Lempel and Ziv (1976), where the measure is introduced and theorems regarding the suitability of the measure are proved. We refer to the measure of complexity introduced by Lempel and Ziv as LZ complexity for brevity.

The LZ complexity measures the number of distinct patterns that must be copied to reproduce a given string. Therefore the only computer operations considered in constructing a string are copying old patterns and inserting new ones. Briefly described, a string $S = s_1 s_2 \ldots s_n$ is scanned from left to right, and a complexity counter $c(S)$ is increased by one unit everytime a new substring of consecutive digits is encountered in the scanning process. The resultant number $c(S)$ is the complexity measure of the string S. Clearly any procedure such as this will over estimate the complexity of strings, but nevertheless we expect comparisons to be meaningful.

In quantifying these ideas it becomes necessary to introduce certain definitions. We make every attempt to keep this to a minimum and omit all proofs of theorems. For these details we refer the reader to Lempel and Ziv (1976). Our efforts here are directed toward outlining a computational algorithm and giving examples of the LZ complexity measure.

We let A denote the alphabet of symbols from which the finite length sequences S are constructed and denote the length of these sequences as $l(S)$. A sequence S with $l(S) = n$ may be written in the form $S = s_1 s_2 \ldots s_n$, where $s_i \in A$. The vocabulary of a sequence S, denoted by $v(S)$, is the set of all substrings of S. For example if $A = \{0,1\}$ and $S = 0010$, then

$$v(S) = \{0, 1, 00, 01, 10, 001, 010, 0010\}.$$

If $Q = q_1 \ldots q_m$ and $R = r_1 \ldots r_n$ are strings, then $S = QR$ denotes the concatenated string $q_1 \ldots q_m r_1 \ldots r_n$ with $l(S) = m + n$. We refer to S as an extension of the string Q. The notation $S\pi$ denotes the string $s_1 \ldots s_{n-1}$ obtained from the string $s_1 \ldots s_n$ by eliminating the last symbol.

The heart of the LZ complexity is a copying process. Using the obvious notation for a substring $S(i,j) = s_i \ldots s_j$, one way that a string S may be extended is concatenation with one of its substrings $W = w_1 \ldots w_{j-i+1} = S(i,j)$, $R = SW$. The element $r_{l(S)+m} = w_m$ is copied from s_{i+m-1}, $m = 1, 2, \ldots, j - i + 1$. The only requirement for an extension $R = SQ$ to be of this type is that Q be an element of $v(SQ\pi) = v(R\pi)$. This follows since $Q \in v(R\pi)$ implies the existence of a positive integer $p \le l(S)$ such that $q_i = r_{p+i-1}$, $i = 1, 2, \ldots, l(Q)$. We begin generating R from S by first copying the known symbol $r_p = s_p$ of S to obtain $q_1 = r_{l(S)+1}$; then we obtain

$q_2 = r_{l(S)+2}$ by copying r_{p+1} (which may still be a symbol of S, or if $p = l(S)$, the first and already known symbol of Q), and so on to the last symbol of Q. We say that an extension $R = SQ$ of S is *reproducible* from S if $Q \in v(R\pi)$. In this process it is clear that requiring $Q \in v(R\pi)$, where the last symbol has been dropped from R, is what ensures that R can be obtained from S by copying alone.

Now we can describe precisely the procedure for computing the LZ complexity measure of a string S. The LZ complexity of a given string S is the number of insertions of new symbols required to construct S, where every attempt is made to construct S by copying alone, i.e., without inserting any new symbols. The process is iterative and the first symbol must always be inserted. Consider a string $S = s_1 s_2 \ldots s_n$ and assume that in the construction of the substring $S(1, r) = s_1 s_2 \ldots s_r \cdot$, the insertion of c symbols from the alphabet A were needed. The centered dot following the symbol s_r denotes that s_r is newly inserted. Then we let $Q = s_{r+1}$ and ask if the extension $R = S(1, r)Q$ is reproducible from $S(1, r)$. If the answer is no, then we must insert $Q = s_{r+1}$ into the string (followed by a dot) and cannot obtain it by simply copying. If the answer is yes, then no new symbol from the alphabet is needed, and we proceed to let $Q = s_{r+1} s_{r+2}$, and then ask the same question: Is the extension $R = S(1, r)Q$ reproducible from $S(1, r)$? In this manner the string $S = s_1 \ldots s_n$ is constructed with each inserted symbol followed by a dot. The LZ complexity $c(S)$ of the string S is the number of inserted dots (plus 1 if the string is not terminated by a dot).

Let us consider some examples to help make the procedure clear. Consider $S = 0000$.

(1) $0 \cdot \cdot$.
(2) $Q = 0, SQ = 0 \cdot 0, v(SQ\pi) = \{0\}$, thus $Q \in v(SQ\pi)$.
(3) $Q = 00, SQ = 0 \cdot 00, v(SQ\pi) = \{0, 00\}$, thus $Q \in v(SQ\pi)$.
(4) $Q = 000, SQ = 0 \cdot 000, v(SQ\pi) = \{0, 00, 000\}$, thus $Q \in v(SQ\pi)$.

Hence $S = 0 \cdot 000$ and thus $c(S) = 1 + 1 = 2$.

Consider $S = 0010$.

(1) $0 \cdot \cdot$.
(2) $Q = 0, SQ = 0 \cdot 0, v(SQ\pi) = \{0\}$, thus $Q \in v(SQ\pi)$.
(3) $Q = 01, SQ = 0 \cdot 01, v(SQ\pi) = \{0, 00\}$, thus $Q \notin v(SQ\pi)$.
(4) $Q = 0, SQ = 0 \cdot 01 \cdot 0, v(SQ\pi) = \{0, 1, 00, 01, 001\}$, thus $Q \in v(SQ\pi)$.

Hence $S = 0 \cdot 01 \cdot 0$ and thus $c(S) = 2 + 1 = 3$.

Consider $S = aabcb$.

(1) $a \cdot$.

(2) $Q = a, SQ = a \cdot a, v(SQ\pi) = \{a\}$, thus $Q \in v(SQ\pi)$.

(3) $Q = ab, SQ = a \cdot ab, v(SQ\pi) = \{a, aa\}$, thus $Q \notin v(SQ\pi)$.

(4) $Q = c, SQ = a \cdot ab \cdot c, v(SQ\pi) = \{a, b, aa, ab\}$, thus $Q \notin v(SQ\pi)$.

(5) $Q = b, SQ = a \cdot ab \cdot c \cdot b, v(SQ\pi) = \{a, b, c, aa, ab, bc, aab, abc, aabc\}$,
    thus $Q \in v(SQ\pi)$.

Hence $S = a \cdot ab \cdot c \cdot b$ and thus $c(S) = 3 + 1 = 4$.

It is clear that the parsing obtained in the LZ complexity measure for a given string S is unique. It is furthermore clear from the first example that $c(S) = 2$ is the minimum value for LZ complexity. As remarked earlier, however, only relative values of $c(S)$ are meaningful and in particular it is the comparison with the $c(S)$ for a random string that is meaningful. To this end we note that Lempel and Ziv (1976) have shown that for a random string of length n, the LZ complexity is given by

$$b(n) = \frac{hn}{\log_K(n)}, \qquad (10.5)$$

where K denotes the number of elements in the alphabet A and h denotes the normalized source entropy. The normalized source entropy is nothing more than the information defined in (9.59), divided by the maximum information obtained when each state is equally probable. In other words

$$h = \frac{-1}{\lg N} \sum_{i=1}^{N} p_i \lg p_i \leq 1. \qquad (10.6)$$

So the procedure is now clear for comparing the LZ complexity of a given string to the LZ complexity of random strings of the same length. The normalized entropy h is determined by determining the probability p_i for each state i. This probability is obtained by counting the occurrences of each symbol in the alphabet and then dividing by the total number of symbols in the string. A common occurrence is that each state, or symbol from the alphabet, is equally probable, in which case $p_i = 1/N$ and $h = 1$. Comparing with the complexity for a random string, we then compute

$$\lim_{n \to \infty} [c(S)/b(n)] \qquad (10.7)$$

for a string with n elements. If the ratio in (10.7) is less than 1, then we can conclude that this is due to a pattern formation in the string S.

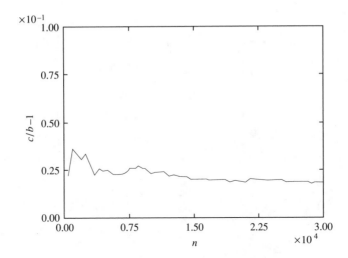

FIGURE 10.1   The approach of the complexity of an n-digit binary string to the theoretical value as a function of the number of digits in the string.

10.3 Complexity and Simple Maps

In this section we consider application of the ideas and results in the previous section to some simple maps we have already studied and we begin by looking at the tent map (2.2).

For the map parameter $\mu = 1$ in (2.2), we know that the tent map iterates are uniformly distributed over the unit interval. Thus we might use a modified tent map as a "random number" generator for this one dimension. We let

$$x_n = 1.0 - 1.99999 \left| (x_{n-1} - 0.5) \right| \tag{10.8}$$

and if $x_n > \frac{1}{2}$, we enter a 1 into the nth slot in the binary string and if $x_n \leq \frac{1}{2}$, we enter a 0 into the nth slot. The factor 1.99999 is used in (10.8) rather than 2.0 because a digital computer always does finite arithmetic, and the "random" number is being generated by truncation at a finite number of digits, rather than by a binary shift on a random initial point with an infinite number of digits. A binary shift on a finite number of digits always leads to 0 after a number of iterations of the map equal to the bit accuracy.

We also note that the approach to the value for a random number in the LZ complexity (10.5) is from above as $n \to \infty$. For example, with eight digits the maximum complexity is $c = 5$ and $b(8) = \frac{8}{3} < 5$. In Fig. 10.1 we see the difference of the ratio in (10.7) from 1.0 plotted as a function of n, the number of digits in the binary string. Employing a standard pseudo-random number generator leads to qualitatively similar results as the modified tent map of (10.8).

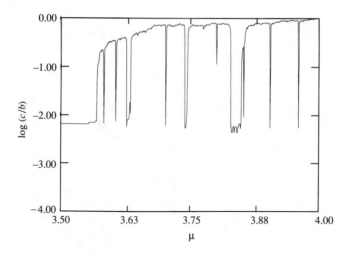

FIGURE 10.2   The LZ complexity of strings constructed from the logisitic map as a function of the map parameter μ.

Figure 10.1 shows the degree to which we can distinguish with confidence the difference between the complexity of a given string and that of a random number with the same number of digits. Based on this figure, we regard 10,000 points on the attractor as sufficient for measuring the complexity of a one-dimensional map and consider the complexity of the attractors generated by logistic maps. Figure 10.2 shows the normalized complexity for the logistic map plotted as a function of the map parameter μ. This figure should be compared with Figs. 2.8 and 4.6. For the logistic map at each value of μ, the digit 0 was entered into the string at the nth position if $x_n \leq \frac{1}{2}$, and the digit 1 was entered if $x_n > \frac{1}{2}$. With this alphabet $\{0, 1\}$, and this manner of assigning digits to the string, the probabilities p_1 and p_2 for the left and right halves of the unit interval are not equal and consequently $h < 1$. An examination of Fig. 2.7 shows that for the smallest value of μ in Fig. 10.2, $p_1 = \frac{1}{4}$ and $p_2 = \frac{3}{4}$ resulting in $h = 0.811$. Only as $\mu \to 4.0$ does $h \to 1.0$ and the complexity approaches that of a random number.

10.4 Summary

In this short chapter algorithmic complexity and the LZ measure of this complexity have been introduced and then applied to the simple dynamical system represented by the logistic map. Kasper and Schuster (1987) have considered other more complex dynamical systems as well and argue for the utility of LZ complexity in detecting order. However, complexity measure at this stage of development seems unable to characterize attractors to the same extent that other measures of fractal sets can. It appears to be a powerful tool for distinguishing dynamical chaos from randomness but not much else. Nevertheless, measures of complexity and their applications to the study of dynamical systems is clearly just beginning, and the potential for added insight into the world from this approach appears significant.

Exercises

10.1 Find the LZ complexity of the string

$$S = 0001101001000101$$

and display the parsing.

10.2 Construct an algorithm for computing the LZ complexity and check it on known patterns.

10.3 Let the assignment of digits in the logistic map study be according to whether or not the nth iterate x_n is less than or equal to the fixed point $x_* = (1 - \mu)/\mu$. Is the LZ complexity the same as that displayed in Fig. 10.2?

10.4 Examine the complexity of the Hénon map (7.40) by establishing an alphabet and a scheme for constructing strings from this alphabet according to where the (x, y) points fall in the plane under iteration of the map.

REPRISE

Having arrived at the conclusion to this text, it is perhaps well to look back on the path traveled—to look back where we have been in order to more clearly see where we are, and have yet to go. We learned much from some simple examples, particularly the logistic map. Based on this one single mapping, a fairly complete picture of the period-doubling route to chaos is known, although not everything was discussed in this book. We examined the quasiperiodic and intermittent transistions to chaos along with some examples, but here again many other examples are known wherein the characteristic signatures of their routes to chaos have been elaborated. We examined for both maps and flows the basic features of normal forms and their associated bifurcations, particularly for the illustrative examples chosen.

Despite the progress that has been made in understanding simple examples, their bifurcations, and their transistions to chaos, we were able to say very little about the manner in which complicated dynamical systems reduce their dynamics to the chaos represented by the simple maps and flows. For example, the task of elucidating the manner in which the Navier-Stokes equations of fluid dynamics reduce to a quadratic map, exhibiting all the features of a period-doubling transistion to chaos, yet remains. We have much to learn about the proper ways to mathematically model physical systems so that experimentally observed chaos is adequately represented.

One important bridge over this difficulty is the seeming universality of chaos. It is extemely exciting to think that vastly different physical systems, through processes yet to be described, can all reduce down to a few universal forms. Universality remains an important and fruitful area of research and promises to contribute to the synthesis and consolidation of science in regards to a large number of complex and seemingly disparate physical phenomena.

We concluded the text by examing some of the more fruitful ways to uncover chaos in experimental data. Fractal dimension and Lyapunov Characteristic Exponents are tried and true methods, but an analysis of the complexity holds promise.

To some extent we have been able to impose order on chaos, not that things are any more predictable, but that our increased understanding has shown us the beauty in chaos—a simplicity in complexity.

GLOSSARY

algorithmic complexity: a measure of the complexity of a data string or program. It is defined to be the length in bits of the shortest algorithm required for a computer to produce the given string or program.

Arnold tongues: resonance zones emanating out from rational numbers in a two-dimensional parameter space of variables.

attractor: a set of points in the space of variables for a dissipative system to which orbits or trajectories tend in the course of dynamic evolution. Fixed points, limit cycles, and strange (chaotic) attractors are examples.

autonomous: refers to dynamical systems where the evolution of dynamical variables depends only on the values of the variables at the previous instant and not on the time explicitly.

basin of attraction: the set of points in the space of system variables such that initial conditions chosen in this set dynamically evolve to a particular attractor.

bifurcation: the sudden appearance of a qualitatively different solution to the equations for a nonlinear system as a parameter is varied.

Cantor set: a closed set of points, completely disconnected, for which every point is a limit point.

chaos: irregular, unpredictable behavior caused by inherent nonlinearites in a dynamical system–not a result of random forces.

cycle: a periodic orbit for a map.

degrees of freedom: the number of independent variables needed to specify the configuration of a system.

doubling transformation: transformation of a function achieved by composing a function with itself that often involves scaling.

Duffing's equation: a differential equation for a nonlinear, driven oscillator where the nonlinear force term is cubic, e.g., $\ddot{x} - \beta x(1 - x^2) = -\alpha \dot{x} + f \cos wt$.

Feigenbaum constants: universal constants for functions approaching chaos via period doubling. The constant δ characterizes the geometric approach of the bifurcation parameter to its limiting value, and the constant α denotes the limiting value for the scaling.

fixed point: a point for the system that is left unchanged by dynamical evolution.

flow: the evolution of points in phase space for a system whose dynamics are represented by a system of differential equations.

focus: a fixed point of a map or differential system that is the focal point of trajectories or orbits that spiral in or out.

fractal dimension: measure of a set of points that characterizes its space-filling properties.

Hénon map: a standard two-dimensional map that exhibits a strange attractor $(x, y) \mapsto (1 - ax^2 + y, bx)$.

Hopf bifurcation: bifurcation of a fixed point to a limit cycle.

horseshoe: characterizes the effect of folding and stretching on a rectangular set induced by some nonlinear maps.

hyperbolic: describes a fixed point for which the linear stability is determined.

information: a binary representation of our knowledge about the state of a system; $I = - \sum_i p_i \log_2 p_i$, where p_i is the probability of the ith state.

instability: : (usually exponential) growth of solutions for dynamical systems.

intermittency: regular or periodic behavior interrupted by chaotic bursts.

invariant manifold: a subset of solutions for a dynamical system that remains invariant under the map or the flow.

KAM theorem: an important theorem in the theory of conservative dynamical systems that ensures the preservation of irrational invariant tori for sufficiently small perturbations.

Koch snowflake: frequently used example of a closed fractal curve with a fractal dimension $\simeq 1.26$.

K-entropy: measure of the rate of change of information.

Lie bracket: the vector field formed from two given vector fields **u:** and **v:** with the jth component given by $\sum_i v^i \partial u^j / \partial x^i - u^i \partial v^j / \partial x^i$.

limit cycle: an attracting set to which orbits or trajectories converge and upon which the dynamics is periodic.

logistic map: a standard one-dimensional map on the interval $[0,1]$, often used to introduce period-doubling bifurcations: $x \mapsto \mu x(1 - x)$.

Lorenz system: a standard system of three coupled differential equations exhibiting chaotic dynamics: $\dot{x} = \sigma(y - x), \dot{y} = \rho x - y - xz, \dot{z} = -\beta z + xy$.

Lyapunov characteristic exponent: a measure of the average exponential separation or contraction of orbits and trajectories.

map, mapping: dynamical rule for obtaining an image point, $\mathbf{x}_n \mapsto \mathbf{x}_{n+1} = \mathbf{f}(\mathbf{x}_n)$.

mode-locking: the nonlinear interaction of a dynamical system to produce periodic behavior that persists for a range of parameters.

nonlinear: describes the response of a dynamical system to a change in some input variable wherein the response is not simply proportional to the change in the variable.

normal form: a standard form for nonlinear differential equations or maps achieved by coordinate transformations.

O-**point**: a fixed point in a conservative system for which nearby points remain nearby under dynamical evolution.

period doubling: bifurcation sequence of periodic orbits for a dynamical system in which the period doubles at each bifurcation.

Poincaré section: a section plane chosen to intersect the trajectories of a dynamical system —useful as an aid in visualizing the motion in complicated systems.

quasiperiodic: describes the motion of a dynamical system containing two incommensurate frequencies.

renormalization: a procedure that leads to form invariance for maps that consists of a combination of translation and rescaling of coordinates.

resonance: refers to the response of a dynamical system with at least two variables in which the periodicities in the variables are commensurate.

self-similar: refers to a point set or system that appears the same under repeated magnifications.

strange attractor: an attracting set that has zero measure in the embedding phase space and has fractal dimension.

subcritical: the property of a bifurcation in which the nonlinear terms do not lead to a stable attractor, i.e., the bifurcated solution is unstable.

supercritical: property of a bifurcation in which the nonlinear terms stabilize the dynamics leading to a stable solution.

supercycle: a periodic cycle for the logistic map with the property that the derivative of the map vanishes at one point of the cycle.

tent map: a piecewise-linear, one-dimensional map on the interval $[0,1]$ exhibiting chaotic dynamics: $x \mapsto \mu(1 - 2|x - \frac{1}{2}|)$.

trajectory: a path followed by a point in the phase space representing the state of a dynamical system described by differential equations.

trapping region: a bounded region from which a point representing the state of a dynamical system can never escape by dynamical evolution.

unimodular function: any function on the interval $[0,1]$ such that the function vanishes at the endpoints and has a single maximum or minimum.

universal function: a function to which all functions that undergo the same sequence of transitions to chaos approach in the limit. A universal function characterizes in a system-independent way properties of a particular kind of chaos.

X-**point**: a fixed point in a conservative dynamical system from which nearby points rapidly leave under dynamical evolution.

REFERENCES

Abramowitz, M., and Stegun, I. A., Eds. (1972) *Handbook of Mathematical Functions* (Dover, New York).

Albano, A. M., Muench, J., Schwartz, C., Mees, A. I., and Rapp, P. E. (1988), *Phys. Rev.* **A38**, 3017.

Amritkar, R. E., Gangal, A. D., and Gupte, N. (1987), *Phys. Rev.* **A36**, 2850.

Arnold, V. I. (1977), *Functional Anal. Appl.* **11**, 85.

Arnold, V. I. (1978), *Mathematical Methods of Classical Mechanics* (Springer-Verlag, New York).

Arnold, V. I. (1983), *Geometrical Methods in the Theory of Ordinary Differential Equations* (Springer, New York).

Arecchi, F. T., and Lisi, F. (1982), *Phys. Rev. Lett.* **49**, 94.

Arecchi, F. T., Meucci, R., Puccioni, G., and Tredicce, J. (1982), *Phys. Rev. Lett.* **49**, 1217.

Atmanspacher, H., and Morfill, G. (1986), *Einführung in die Theorie das Deterministischen Chaos*, MPE Report 196, (Max-Planck- Institut, Garching bei München).

Atmanspacher, H., and Scheingraber, H. (1986), *Phys. Rev.* **A35**, 253.

Bak, P. (1986), *Physics Today* **39** (Dec.), 41.

Belmonte, A. L., Vinson, M. J., Glazier, J. A., Gemunu, H. G., and Kenny, B. G. (1988), *Phys. Rev. Lett.* **61**, 539.

Beltrami, E. (1987), *Mathematics for Dynamic Modeling* (Academic Press, Boston).

Benettin, G., and Galgani, L. (1979), in *Intrinsic Stochasticity in Plasmas*, G. Laval and D. Gresillon, Eds. (Institut. d'etudes scientifiques de Cargise, Corse), 93.

Benettin, G., Galgani, L., and Strelcyn, J.-M. (1976), *Phys. Rev.* **A14**, 2338.

Benettin, G., Galgani, L., Giorgilli, A., and Strelcyn, J.-M. (1980), *Meccanica* **9**, 9.

Berge, P., Pomeau, Y., and Vidal, C. (1984), *Order Within Chaos* (Wiley, New York).

Bohr, T., Bak, P., and Jensen, M.H. (1984), *Phys. Rev.* **A30**, 1970.

Brandstater, A., and Swinney, H. L. (1987), *Phys. Rev.* **A35**, 2207.

Braun, M. (1983), *Differential Equations and Their Applications* (Springer, New York).

Briggs, K. (1987), *Am. J. Phys.* **55**, 1083.

Bullard, E. (1978), in *AIP Conference Proceedings*, S. Jorna, Ed. (AIP, New York), Vol. 46, 373.

Byrd, P. F., and Friedman, M. D. (1954), *Handbook of Elliptic Integrals for Engineers and Physicists* (Springer, New York).

Chaitin, G. J. (1974), *IEEE Transac. Inform. Theory* **IT-20**, 10.

Chaitin, G. J. (1975), *Sci. American* **232**, 47.

Chen, C., Gyorgyi, G., and Schmidt, G. (1987), *Phys. Rev.* **35A**, 2660.

Chernikov, A. A., Sagdeev, R. Z., Usikov, D. A., Zakharov, M. Yu, and Zaslavsky, G. M. (1987), *Nature* **326**, 559.

Chirikov, B. V. (1979), *Phys. Reps.* **52**, 265.

Cheung, P. Y., and Wong, A. Y. (1987), *Phys. Rev. Lett.* **59**, 551.

Cheung, P. Y., Donovan, S., and Wong, A. Y. (1988), *Phys. Rev. Lett.* **61**, 1360.

Chow, S.-N., and Hale, J. K. (1982), *Methods of Bifurcation Theory* (Springer, New York).

Collet, P., Eckman, J. P., and Lanford, D. E. (1980), *Commun. Math. Physics* **76**, 211.

Collet, P., Eckmann, J. P., and Koch, H. (1981), *Physica* **3D**, 457.

Cook, A. E., and Roberts, P. H. (1970), *Proc. Camb. Phil. Soc.* **68**, 547.

Cvitanović, P. (1984), *Universality in Chaos* (Adam Hilger, Bristol).

Cvitanović, P., Jensen, M. H., Kadanoff, L. P., and Procaccia, I. (1985), *Phys. Rev. Lett.* **55**, 343.

D'Humieres, D., Beasley, M. R., Huberman, B. A., and Liebenhaber, A. (1982), *Phys. Rev.* **26A**, 3483.

Derrida, B., Gervois, A., and Pomeau, Y. (1979), *J. Phys.* **12A**, 269.

Devaney, R. L. (1987), *An Introduction to Chaotic Dynamical Systems* (Addison-Wesley, New York).

Epstein, I. R. (1983), in *Order in Chaos*, D. Campbell and H. Rose, Eds. (North-Holland, Amsterdam), 47.

Fante, R. L. (1988), *Signal Analysis and Estimation: An Introduction* (Wiley, New York).

Farmer, J. D. (1982), in *Evolution of Order and Chaos*, H. Haken, Ed. (Springer, New York), 228.

Farmer, J. D., Ott, E., and Yorke, J. A. (1983), in *Order in Chaos*, D. Campbell and H. Rose, Eds. (North-Holland, Amsterdam), 53.

Feigenbaum, M. J. (1979), *J. Stat. Physics* **21**, 669.

Feigenbaum, M. J. (1980a), *Los Alamos Science* **1**; reprinted in *Universality in Chaos*, P. Cvitanovic, Ed. (Adam Hilger, Bristol, 1984).

Feigenbaum, M. J. (1980b), *Commun. Math. Phys.* **77**, 65.

Feigenbaum, M. J., Kadanoff, L. P., and Shenker, S. J. (1982), *Physica* **5D**, 370.

Ford, J. (1975), in *Fundamental Problems in Statistical Mechanics*, E. G. D. Cohen, Ed. (North Holland, Amsterdam).

Ford, J. (1986), in *Chaotic Dynamics and Fractals*, M. F. Barnsley and S. G. Demko, Eds. (Academic, New York), 1.

Ford, J. (1988), in *Directions in Chaos*, H. Bai-lin , Ed. (World Scientific, New Jersey), Vol. 2, 128.

Frederickson, P., Kaplan, J., Yorke, E., and York, J. (1983), *J. Diff. Eqns.* **49**, 185.

Gibbs, H. M., Hopf, H. A., Kaplan, D. L., and Shoemaker, R. L. (1981), *Phys. Rev. Lett.* **45**, 709.

Giglio, M., Masazzi, S., and Perini, V. (1981), *Phys. Rev. Lett.* **47**, 243.

Glass, L. Guevara, M. R., and Shrier, A. (1983), in *Order in Chaos*, D. Campbell and H. Rose, Eds. (North-Holland, Amsterdam), 89.

Goldstein, H. (1980), *Classical Mechanics* (Addison-Wesley, Reading, Mass.).

Gollub, J. P., and Benson, S. V. (1980), *J. Fluid Mech.* **100**, 449.

Grassberger, P. (1981), *J. Stat. Phys.* **26**, 173.

Grassberger, P. (1983), *Phys. Lett.* **97A**, 224.

Grassberger, P. (1985), in *Chaos in Astrophysics*, J.R. Buchler, J.M. Perdang, and E.A. Spiegel, Eds. (D. Reidel, Dordrecht), 193.

Grassberger, P., and Procaccia, I. (1983a), *Physica* **9D**, 189.

Grassberger, P., and Procaccia, I. (1983b), *Phys. Rev. Lett.* **50**, 346.

Greene, J. M. (1979), *J. Math. Phys.* **20**, 1183.

Greene, J. M., Mackay, R. S., Vivaldi, F., and Feigenbaum, M. J. (1981), *Physica* **3D**, 468.

Guckenheimer, J., and Holmes, P. (1983), *Nonlinear Oscillators, Dynamical Systems, and Bifurcations of Vector Fields* (Springer, New York).

Halsey, T. C., Jenson, M. H., Kadanoff, L. P., Procaccia, I., and Shraiman, B. I. (1986), *Phys. Rev.* **A33**, 1141.

Hammel, S. M., Yorke, J. A., and Grebogi, C. (1988), *Bull. Amer. Math. Soc.* **19**, 465.

Harth, E. (1983) *IEEE Transactions* **SMG-13**, 782.

Helleman, R. H. G. (1980), in *Fundamental Problems in Statistical Mechanics V*, E.G.D. Cohen, Ed. (North Holland, Amsterdam), 165.

Hénon, M. (1969), *Quart. Appl. Math.* **27**, 291.

Hénon, M. (1976), *Comm. Math. Phys.* **50**, 69.

Hénon, M., and Heiles, C. (1964), *Astron. J.* **69**, 73.

Hentschel, H. G. E., and Procaccia, I. (1983), *Physica* **8D**, 435.

Hirsch, J. E., Huberman, B. A., and Scalapino, D. J. (1982), *Phys. Rev.* **25A**, 519.

Holmes, P. J., and Whitley, D. (1983), in *Order in Chaos*, D. Campbell and H. Rose, Eds. (North-Holland, Amsterdam), 111.

Hopf, F. A., Kaplan, D. L., Gibbs, H. M., and Shoemaker, R. L. (1982), *Phys. Rev.* **A25**, 2172.

Hu, B. (1982), *Phys. Reps.* **91**, 233.

Hu, B., and Rudnick, J. (1982), *Phys. Rev. Lett.* **48**, 1645.

Huberman, B. A., Crutchfield, J. P., and Packard, N. H. (1980), *Appl. Phys. Lett.* **37**, 750.

Hurd, A. J. (1988), *Am. J. Phys.* **56**, 969.

Ikeda, K., Daido, H., and Akimoto, O. (1980), *Phys. Rev. Letters* **45**, 709.

Iooss, G. (1979), *Bifurcations of Maps and Applications* (North-Holland, New York).

Irwin, M. C. (1980), *Smooth Dynamical Systems* (Academic Press, New York).

Jensen, M. H., Bak, P., and Bohr, T. (1983), *Phys. Rev. Lett.* **50**, 1637.

Jensen, M. H., Bak, P., and Bohr, T. (1984), *Phys. Rev.* **30A**, 1970.

Kadanoff L. P. (1985), in *Regular and Chaotic Motions in Dynamic Systems*, G. Valo and A. S. Wightman, Eds., NATO ASI Series B: Physics, Vol. 118 (Plenum, New York).

Kaplan, J., and Yorke, J. (1978), in *Functional Differential Equations and the Approximation of Fixed Points*, Lecture Notes in Mathematics 730, H. O. Peitgen and H. O. Walther, Eds. (Springer, New York), 204.

Kaspar, F., and Schuster, H. G. (1987), *Phys. Rev.* **A36**, 842.

Kobayashi, S. (1962), *Trans. Japan Society Aeronautical Space Sciences* **5**, 90.

Kubicek, M., and Marek, M. (1983), *Computational Methods in Bifurcation Theory and Dissipative Structures* (Springer, New York).

LaBrecque, M. (1986), *Mosaic* **17**, 34.

LaBrecque, M. (1987), *Mosaic* **17**, 23.

Lanford, D. E. (1982), *Bull. Am. Math. Soc.* **6**, 427.

Lauterborn, W., and Cramer, E. (1981), *Phys. Rev. Lett.* **47**, 1445.

Lauwerier, H. A. (1986), in *Chaos*, A. V. Holden, Ed. (Princeton University Press, Princeton), part II.

Lempel, A., and Ziv, J. (1976), *IEEE Transac. Inform. Theory* **IT-22**, 75.

Lewin, R. (1988), *Science* **240**, 986.

Libchaber, A., and Maurer, J. (1978), *J. Phys. Lett.* **39**, 369.

Libchaber, A., and Maurer, J. (1981), in *Nonlinear Phenomena at Phase Transitions and Instabilities*, T. Riste, Ed. (Plenum, New York), 259.

Libchaber, A., Laroche, C., and Fauve, S. (1982), *J. de Phys. Lett.* **43**, L211.

Lichtenberg, A. J., and Lieberman, M. A. (1983), *Regular and Stochastic Motion* (Springer, New York).

Linsay, P. S. (1981), *Phys. Rev. Lett.* **47**, 1349.

Lorenz, E. N. (1963), *J. Atmos. Sci.* **20**, 130.

Majumdar, S., and Prasad, B. R. (1988), *Comp. Phys.* **2**, 69.

Mandelbrot, B. B. (1983), *The Fractal Geometry of Nature* (Freeman, San Francisco).

Mandelbrot, B. B., Passoja, D. E., and Paullay, A. J. (1984), *Nature* **308**, 721.

Mao, J., and Greene, J. M. (1987), *Phys. Rev.* **35A**, 3911.

Marsden, J. E., and McCracken, M. (1976), *The Hopf Bifurcation and Its Applications* (Springer, New York).

Martens, C. C., Davis, M. J., and Ezra, G. S. (1987), *Chem. Phys. Lett.* **142**, 519.

May, R. M. (1976), *Nature* **261**, 459.

Moon, F. C. (1987), *Chaotic Vibrations* (Wiley, New York).

Moon, F. C., and Holmes, P. J. (1979), *J. Sound Vib.* **65**, 285.

Moon, F. C., and Holmes, P. J. (1980) *J. Sound Vib.* **69**, 339.

Moon, F. C., and Li, G.–X. (1985), *Phys. Rev. Lett.* **55**,1439.

Nicolis, J. S. (1984), *J. Franklin Instit.* **317**, 289.

Oseledec, V. I. (1968), *Trans. Moscow Math. Soc.* **19**, 197.

Packard, N. H., Crutchfield, J. P., Farmer, J. D., and Shaw, R. S. (1980), *Phys. Rev. Lett.* **45**, 712.

Pedersen, N. F., and Davidson, A. (1981), *Appl. Phys. Lett.* **39**, 830.

Press, W. H., Flannery, B. P., Teukolsky, S. A., and Vetterling, W. T. (1986), *Numerical Recipes* (Cambridge, New York).

Pomeau, Y., and Mannville, P. (1980), *Comm. Math. Phys.* **74**, 189.

Rasband, S .N. (1983), *Dynamics* (Wiley, New York).

Roux, J.–C. (1983), in *Order in Chaos*, D. Campbell and H. Rose, Eds. (North-Holland, Amsterdam), 57.

Russell, D. A., Hanson, J. D., and Ott, E. (1980), *Phys. Rev. Lett.* **45**, 1175.

Sarkar, S. K. (1987), *Phys. Rev.* **A36**, 4104.

Sattinger, D. H. (1979), *Group Theoretical Methods in Bifurcation Theory* (Springer, New York).

Schlögl, F. (1980), *Phys. Rep.* **62**, 267.

Schuster, H. G. (1984), *Deterministic Chaos* (Physik-Verlag, Weinheim, FRG).

Schutz, B. F. (1980), *Geometrical Methods of Mathematical Physics* (Cambridge University Press, Cambridge).

Shannon, C. E., and Weaver, W. (1949), *The Mathematical Theory of Information* (University of Illinois Press, Urbana).

Skarda, C. A., and Freeman, W. J. (1987), *Behav. Brain Sci.* **10**, 161.

Shimada, I., and Nagashima, T. (1979), *Prog. Theor. Phys.* **61** 1605.

Simoyi, R. H., Wolf, A., and Swinney, H. L. (1982), *Phys. Rev. Lett.* **49**, 245.

Smith, C. W., Tejwani, M. J., and Farris, D. A. (1982), *Phys. Rev. Lett.* **48**, 492.

Smith, K. T. (1983), *Primer of Modern Analysis* (Springer, New York), 245.

Sparrow, C. (1982), *The Lorenz Equations: Bifurcations, Chaos, and Strange Attractors* (Springer, New York).

Swinney, H. L. (1983), in *Order in Chaos*, D. Campbell and H. Rose, Eds. (North-Holland, Amsterdam), 3.

Swinney, H. L. (1985), in *Fundamental Problems in Statistical Mechanics VI*, E. G. D. Cohen, Ed. (Elsevier Science Publishers, New York), 253.

Swinney, H. L., and Gollub, J. P. (1978), *Physics Today* **31**, 41 (August).

Takens, F. (1981), in *Lecture Notes in Mathematics*, D. A. Rand and L. S. Young, Eds. (Springer, New York), 366.

Testa, J. S., Pérez, J., and Jeffries, C. (1982), *Phys. Rev. Lett.* **48**, 714.

Van der Pol, B., and Van der Mark, J. (1927), *Nature* **120**, 363.

Weiss, C. O., Godone, A., and Olafsson, A. (1983), *Phys. Rev.* **A28**, 892.

West, B. J., and Goldberger, A. L. (1987), *Amer. Scientist* **75**, 354.

Whitley, D. (1983), *Bull. London Math. Soc.* **15**, 177.

Yeh, W. J., and Kao, Y. J. (1982), *Phys. Rev. Lett.* **49**, 1888.

INDEX